Recent Updates in HVAC Systems

Edited by César Martín-Gómez

Published in London, United Kingdom

Recent Updates in HVAC Systems
http://dx.doi.org/10.5772/intechopen.104038
Edited by César Martín-Gómez

Contributors
Sami Missaoui, Zied Driss, Romdhane Ben Slama, Bechir Chaouachi, Aung Myat, Nan Chen, Yunshui Chen, He Zhao, Igor Zhadanovsky, Muhammad Ahmad Jamil, Muhammad Wakil Shahzad, Kim Choon Ng, Haseeb Yaqoob, Muhammad Waqar Ashraf, Nida Imtiaz, Ben Bin Xu

First published in London, United Kingdom, 2023 by IntechOpen
IntechOpen is the global imprint of INTECHOPEN LIMITED, registered in England and Wales, registration number: 11086078, 5 Princes Gate Court, London, SW7 2QJ, United Kingdom

British Library Cataloguing-in-Publication Data
A catalogue record for this book is available from the British Library

Additional hard and PDF copies can be obtained from orders@intechopen.com

Recent Updates in HVAC Systems
Edited by César Martín-Gómez
p. cm.
Print ISBN 978-1-83768-173-0
Online ISBN 978-1-83768-174-7
eBook (PDF) ISBN 978-1-83768-175-4

Meet the editor

César Martín Gómez, Ph.D., is an architect responsible for building services in complex buildings such as the Auditorium of Navarra, the V Stage of Universidad de Navarra Clinic, and the Spanish Pavilion at the Saragossa Expo. He has worked at I&S Ingenieros, in the Architecture Department of the Spanish Renewable Energies Center (CENER), and as a building services and energy coordinator at Mangado & Asociados. Dr. Gómez has worked as a researcher and professor in the Construction, Building Services and Structures Department, Universidad de Navarra, Spain, since 2009.

Contents

Preface

This book presents research on energy self-sufficiency for buildings, focusing on HVAC solutions and their energy systems.

As technology continues to advance, the construction of buildings has become increasingly complex and technologically advanced, similar to the automobile and aeronautics industries. In light of the current COVID-19 pandemic, there is a pressing need for novel solutions to address the escalating technological demands in buildings. This leads to the question of whether traditional HVAC systems will suffice or if alternative methods must be considered to tackle these new challenges.

The near future is set to be revolutionized by disruptive technologies such as hydrogen fuel cells, polygeneration of energy, second-use electric batteries for energy storage in buildings, and 3D printing being extensively employed in construction. As a result, it is crucial to contemplate how HVAC system managers will adapt to these innovative advancements.

The book starts by examining research on waste heat recovery from heat pumps. Experimental data is used to validate the proposed model, which shows that waste heat can be utilized for water heating without disrupting the heat pump cycle, making it suitable for bathing and cleaning. Chapter 2 discusses the application of artificial intelligence in air conditioning systems to resolve typical office issues, showing significant energy savings. Chapter 3 discusses the challenges of reclaiming thermal energy from cryptocurrency mining. It presents an innovative heat recovery system and demonstrates its results. Chapter 4 proposes a new heating solution for retrofitting existing steam heating systems and for new installations in high-rise buildings. The book concludes with Chapter 5, which discusses an indirect evaporative cooler that uses a unique operational setup, making it adaptable to various weather conditions.

César Martín-Gómez
Department of Construction,
Building Services and Structures,
University of Navarra,
Pamplona, Spain

Section 1

Heat Recovery

Chapter 1

Numerical and Experimental Study of the Heat Pump Water Heater with an Immersed Helical Coil Heat Exchanger

Sami Missaoui, Zied Driss, Romdhane Ben Slama and Bechir Chaouachi

Abstract

Traditional water heating and cooling methods consume a significant amount of energy. Heating and cooling were also significant contributors to CO_2 emissions. Based on several previous studies, heat pumps are an economical solution because they reduce the energy consumed for water heating and the carbon dioxide emissions compared to conventional electric resistant water heaters, gas boilers, and so on. As a result, this chapter presented research studies on waste heat recovery from heat pumps' immersed helical condenser coils for domestic hot water production. In this work, the three-dimensional geometry of the storage tank with immersed helical pipe heat exchanger was simplified to two-dimensional axisymmetric geometry and the numerical investigation was conducted with using computational fluid dynamics (CFD) commercial software. The variable heat flux from the condenser coil was considered to heat the water in the tank under laminar flow regime. The proposed model was validated with the experimental data and agreed very well. According to the obtained results, the outlet heat flux tended to the decrease with an increase of the heating time. Also, the outcomes indicate that the waste heat can be used for water heating utilized for bathing and cleaning without affecting the heat pumps cycle.

Keywords: heat pumps, heating, cooling, thermodynamic parameters, heat recovery, helically coiled pipe

1. Introduction

Energy consumption for domestic hot water production is on the rise in both residential and commercial buildings nowadays. Water heaters have traditionally been equipped with traditional heaters that generate heat by consuming fossil fuels, electricity, wood and oil. All of which are undesirable in terms of greenhouse gas emissions and energy consumption. As a result, the coupling of a heat pump system with a water tank is currently a viable recommended approach for water heating as a substitute. Because of its excellent energy efficiency, simple operation, low cost, and environmental friendliness, many academics have been interested in heat pump water heaters (HPWH). Missaoui et al. [1] numerically investigate the HPWH with helically coiled tube heat exchanger. The results indicated that when the copper coil pitch is 20 mm, the water velocity increases and reaches its maximum in the middle part of the tank. Qiang and Shuhong et al. [2] carried out a numerical investigation on the performance of HPWH with wrap-around condenser coil. The results indicated that the heat transfer coefficient of variable pitch coil was increased by 21.91% compared with constant-pitch coil. Missaoui et al. [3] conducted an experimental and numerical analysis of a helical coil heat exchanger for domestic refrigerator and water heating. The results indicated that after 5 hours of heating the average coefficient of performance of the refrigerator domestic coupled with water heating process, the variable pitch coil was increased by 16.17% compared with normal coil. Zhou et al. [4] conducted a study on geometric parameter of a wrap-around condenser for a water tank with 200 L capacity. The results indicated that the pipe parameters such as location, turns and spacing have direct effects on the performance of the HPWH. Sami et al. [5] experimentally validated the mathematical model of the domestic refrigerator for water heating. The results indicated that the domestic refrigerator can be used for hot water production without removing its main role.

The goal of this project is to create a device that can take use of the waste heat produced by refrigeration equipment and use it for home functions such as bathing, laundry, and cleaning.

2. System description

A schematic diagram for the water heating system is shown in **Figure 1**. The air conditioning with domestic hot water production is mainly composed of a water heating system and vapor compression cycle, including compressor, water tank with immersed condenser, an expansion valve and evaporator. This system is based on the same principle of vapor compression cycle. But, there is a small change in the condenser. The conventional heat pump was modified to the air conditioner with a domestic hot water supply that is operates in the domestic hot water production and space-cooling. The helically condenser coil was immersed in the water tank of 80 L capacity.

3. Numerical model

3.1 CFD model

As shown in **Figure 2**, the three dimensional water tank model was simplified as an axisymmetric two-dimensional model.

3.2 Boundary conditions

The boundary conditions of the heat transfer between the water and the condenser is illustrated in **Table 1**.

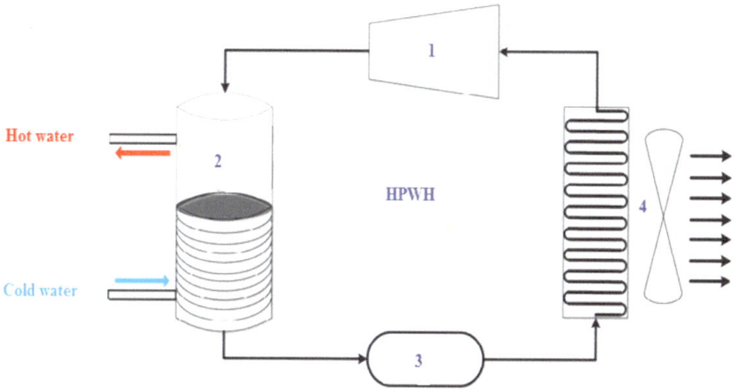

N°	Components
1	Compressor
2	Storage tank with immersed condenser
3	Expansion valve
4	Evaporator

Figure 1.
Schematic diagram of HPWH system.

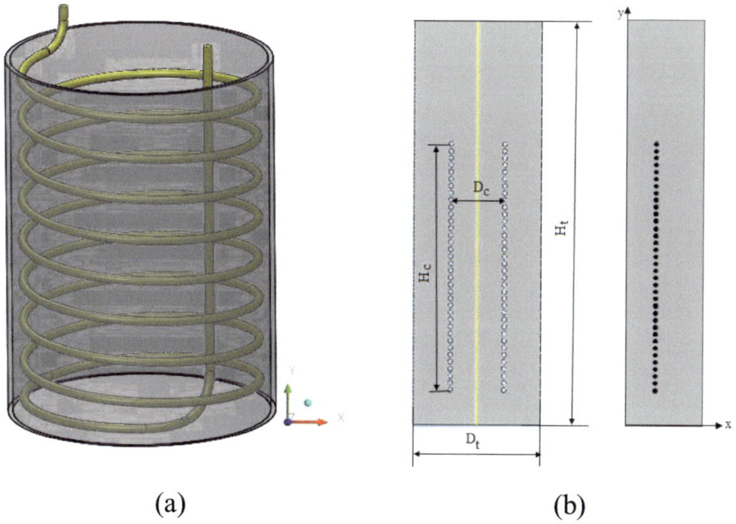

Figure 2.
Simplified model. (a) 3D view. (b) 2D view.

Domain	Specification	Material type	Value
Centerline	Axis	—	—
Tank bottom	Heat flux	Aluminum	0 W/m^2
Tank top	Heat flux	Aluminum	0 W/m^2
Tank wall	Heat flux	Aluminum	0 W/m^2
Cylinder heater	Heater	Copper	Variable

Table 1.
Boundary conditions.

The considered equation was proposed by Dai et al. [6] and it is expressed as follows:

$$q(t) = -0.000000003t^3 + 0.000032972t^2 - 0.18825t + 4376.4 \quad (1)$$

3.3 Meshing

As shown in **Figure 3**, the grid near the condenser coil and the tank wall was refined. The mesh was conducted using the commercial computational fluid dynamics

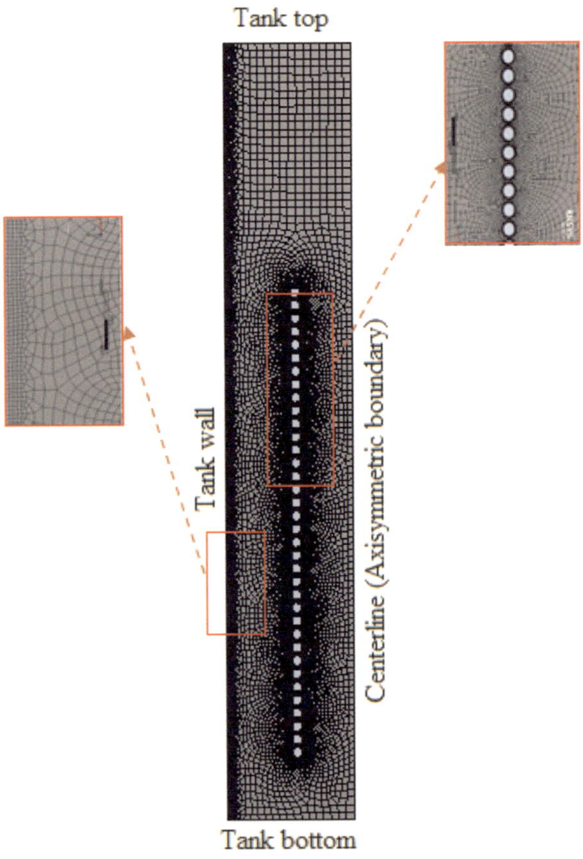

Figure 3.
Mesh and boundary conditions.

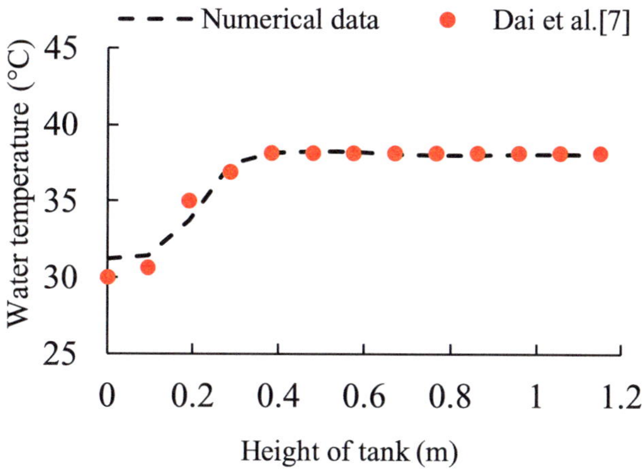

Figure 4.
Model validation results.

(CFD) package Fluent. In the numerical investigation, the mesh with 40105 cells was adopted for further analysis.

4. Model validation

In order to validate the accuracy of our numerical method, we have compared our results to those reported by Dai et al. [7]. From these results, it has been observed that the numerical results are in good agreement with the experimental results referring to a HPWH with immersed condenser coil. In these conditions, the maximum absolute error is about 10% as shown in **Figure 4**.

5. Results and discussion

To calculate the water temperature distribution in each axial location along the height of the tank, the water tank was divided into three layers in the axial direction. Each layer was divided into three points in the radial direction designed by A, B and C. Details for the locations of the different test points are shown in **Figure 5**. In order to calculate the average water temperature in each layer, the Eq. (2) is used:

$$T_{\mathrm{w,y}}(t) = \frac{\sum_{x=1}^{3} \mathrm{T}_{w,x}(t)}{3} \tag{2}$$

Where $T_{w,x}(t)$ is the water temperature in each test point.

Figure 6 shows the average water temperature rising curves of A, B and C in water heating process at three different layers. From these results, it has been observed that the water temperature in the middle and the upper part of the tank was the same. On the other hand, the A water temperature curve rising slowly with slight fluctuations indicating that the hot water in the lower part of the storage tank rises up. This fact is due to the force of gravity (g) acting on the fluid density variations. The majority of water liquid in the lower part of storage tank still has not been affected by the

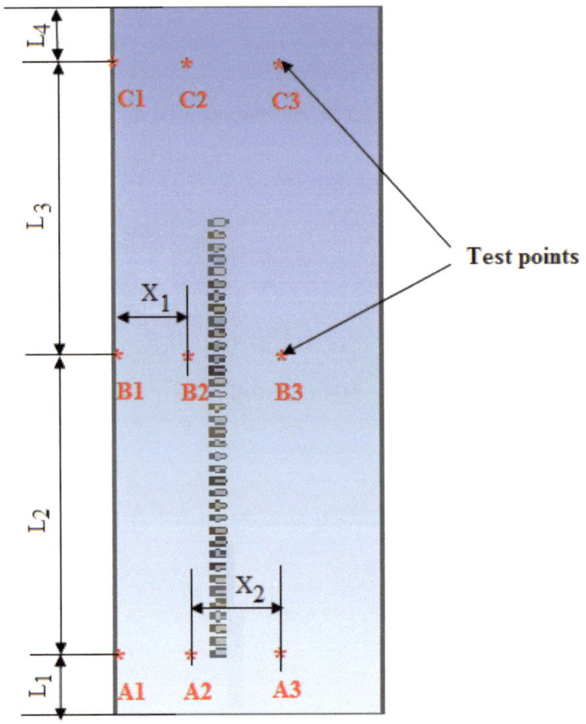

Parameter	Value (mm)
L1	100
L2	475
L3	475
L4	100
X1	60
X2	60

Figure 5.
Locations of different test points.

convective heat transfer, except at the middle and upper part. Thus, this increase in water temperature in the middle and upper part of the storage tank is due to the buoyancy driven flow.

Figure 7 shows the heat flux distribution of the helical coiled tube heat exchanger. From these results, it is clear that the heat flux decreases with the increase of the

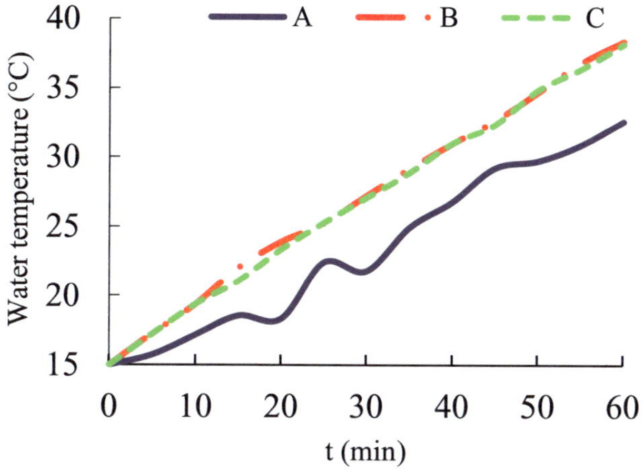

Figure 6.
Water temperature distribution in the water tank.

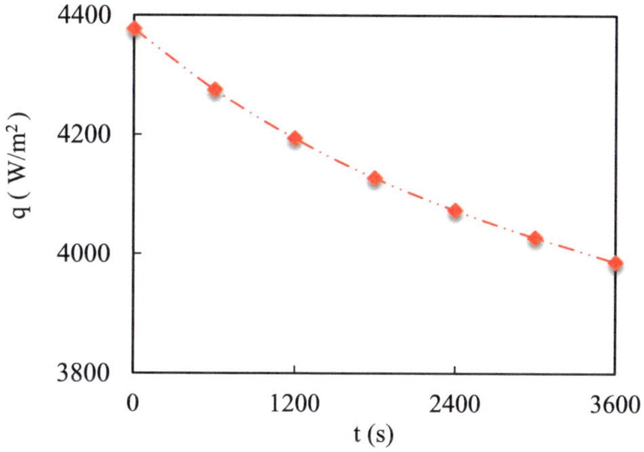

Figure 7.
Heat flux distribution during heating process.

heating time. Therefore, this result declined to the low convective heat transfers between the water in the tank and the refrigerant in the condenser when the water temperature rises up.

Figure 8 shows the distribution of the water velocity over 60 min with the initial water temperature equal to 15°C. The zones near the condenser coil and tank wall are zoomed to obtain more detailed information about the water recirculation during heating process. From these results, it has been observed that the velocity distribution is high in the centerline of the vertical direction of the water tank. The high velocity field is accurately visualized near the tank walls and condenser coil. Indeed, a marked higher velocity distribution has been observed in the middle part of the storage tank. From the zoomed regions, the water near the tank wall and condenser coil flowing downward during the heating process. This fact is due to the water recirculation.

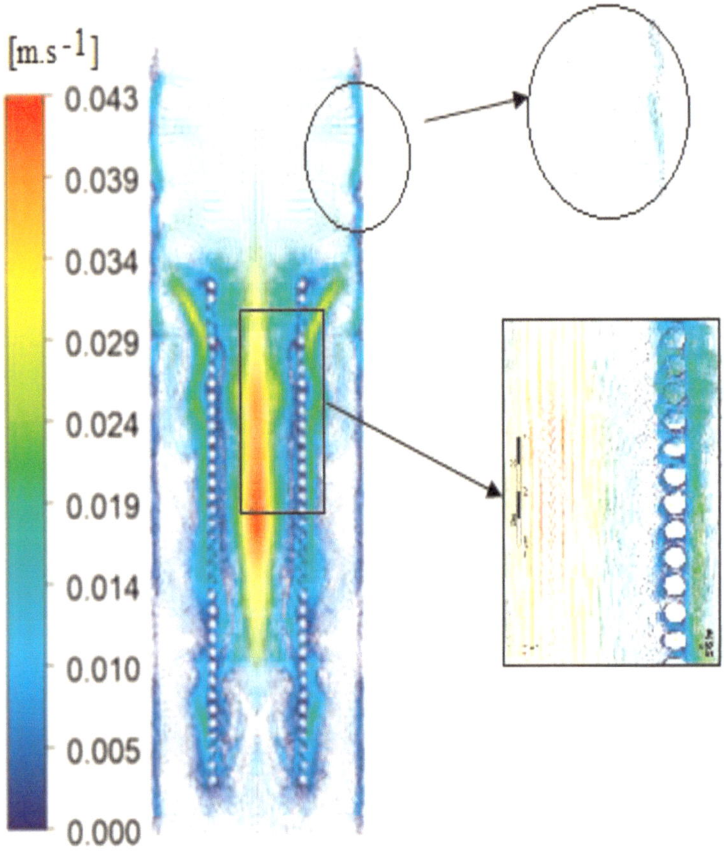

Figure 8.
Velocity distribution of the water in the storage tank at t = *60 min.*

6. Conclusions

In this paper, numerical investigations were carried out to study the heat transfer, water temperature and water velocity distribution in the cylindrical tank during heating process. The detailed outcomes are as follows:

- The maximum water temperature achieved in the cylindrical tank after 3 h of heating is equal to 50°C. This hot water obtained can be used for domestic purposes such as bathing, laundry and cleaning.

- The new coupled machine has good utilization in industry, hotels and also for domestic purpose.

- This work is extremely important for the development of novel heat recovery technologies from refrigeration units in order to obtain cooling at a low energy cost, with no negative environmental impact and a low initial cost.

Acknowledgements

The authors are grateful to the members of the Laboratory of Energy, Water, Environment, and Processes (LR18ES35) for their financial support.

Conflict of interest

The authors declare that they have no known competing financial interests or personal relationships that could have appeared to influence the work reported in this paper.

Author details

Sami Missaoui[1,2]*, Zied Driss[3], Romdhane Ben Slama[2] and Bechir Chaouachi[2]

1 Higher National Engineering School of Tunis (ENSIT), University of Tunis, Tunis, Tunisia

2 Laboratory of Energy, Water, Environment and Processes, National Engineering School of Gabes, University of Gabes, Gabes, Tunisia

3 Laboratory of Electromechanical Systems (LASEM), National School of Engineers of Sfax, University of Sfax, Sfax, Tunisia

*Address all correspondence to: missaouisami1988@gmail.com

References

[1] Missaoui S, Driss Z, Slama RB, Chaouachi B. Numerical analysis of the heat pump water heater with immersed helically coiled tubes. Journal of Energy Storage. 2021;**39**:102547

[2] Ye Q, Li S. Investigation on the performance and optimization of heat pump water heater with wrap-around condenser coil. International Journal of Heat and Mass Transfer. 2019;**143**: 118556

[3] Missaoui S, Driss Z, Slama RB, Chaouachi B. Experimental and numerical analysis of a helical coil heat exchanger for domestic refrigerator and water heating. International Journal of Refrigeration. 2022;**133**:276-288

[4] Zhou H, Chen Y, Luo M, Zhao H, Zhong F, Naiping G. Performance analysis and optimization of wrap-around condenser in an air source heat pump water heater system: Numerical and experimental investigation. In: Proceedings of 12th IEA Heat Pump Conference in Rotterdam. 2017

[5] Missaoui S, Driss Z, Slama RB, Chaouachi B. Experimentally validated model of a domestic refrigerator with an immersed condenser coil for water heating. International Journal of Air-Conditioning and Refrigeration. 2021; **29**(3):2150022 (13 pp)

[6] Dai NN, Li SH. Simulation and performance analysis on condenser coil in household heat pump water heater. Sustainable Cities and Society. 2018;**36**: 176-184

[7] Dai NN, Li SH, Ye Q. Performance analysis on the charging and discharging process of a household heat pump water heater. International Journal of Refrigeration. 2019;**98**:266-273

Chapter 2

Application of Artificial Intelligence in Air Conditioning Systems

Aung Myat

Abstract

Urbanization has led to a sharp rise in the demand for power over the past 10 years, alarmingly rising greenhouse gas (GHG) emissions. HVAC (heating, ventilation, and air conditioning) systems account for nearly half of the energy used by buildings, and minimizing the energy use of the HVAC systems is essential. However, the common problems, such as hot spots and cold spots in office spaces, experienced in the building need to be addressed. Therefore, this chapter introduces the application of artificial intelligence proactive control to resolve typical office issues. A demonstration testbed was implemented on the Singapore Institute of Technology (SIT) campus. The experiments were conducted in baseline mode and smart mode. In the case study, two big zones were segregated into 43 micro-zones equipped with smart dampers at each diffuser, allowing a localized set point to improve thermal comfort and eliminate hot and cold spots. It has been observed that the proactive AI control reduces cooling provided to the office by 29 percent and AHU electricity usage by 50 percent, respectively, while keeping the area within thermal comfort range of 23 to 25°C and 50 to 63% relative humidity.

Keywords: energy efficiency, all-air systems, airside energy reduction, artificial intelligence, micro-zones concept, energy savings

1. Introduction

Energy is the most important component for the operation of various sectors, including transportation, business, residential buildings, and many others. Recent technological developments have led to a sharp rise in global energy consumption, which is alarmingly increasing the rate of greenhouse gas emissions. As shown in **Figure 1**, the world energy consumption by different sources of fuels was about 173,340 Terra-Watt-Hr (TWh) in 2019, while it was 122,073 TWh in 2000. The world's energy consumption increased by approximately 42% within 19 years. Electricity is the prime energy source that the built environment utilizes. Global electricity generation in 2021 increases approximately twofold compared to 2000 to accommodate the drastic increase in energy consumption in the built environment, as indicated in **Figure 2**. Primary fuel sources, like coal and gas, account for almost 60% of total primary energy sources, whereas renewable energy makes up only 13%

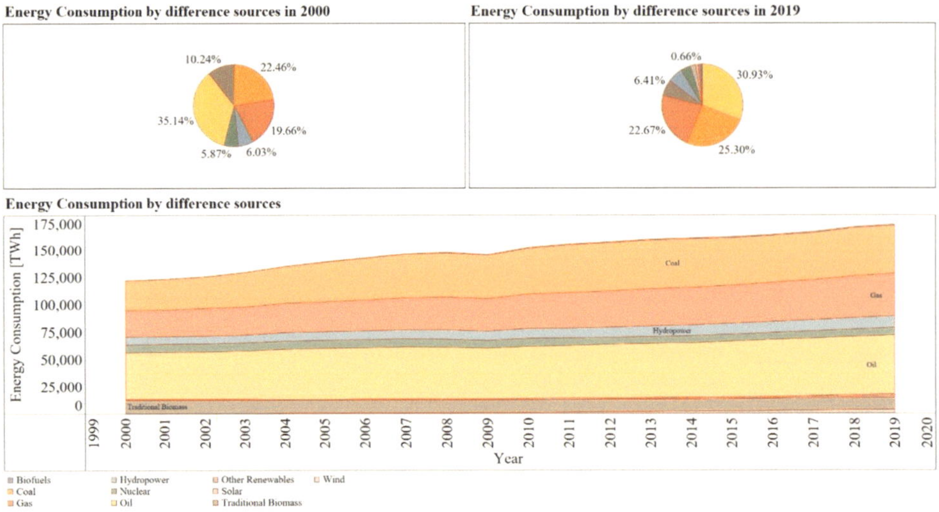

Figure 1.
Energy consumption by different fuel sources since 2000 [1].

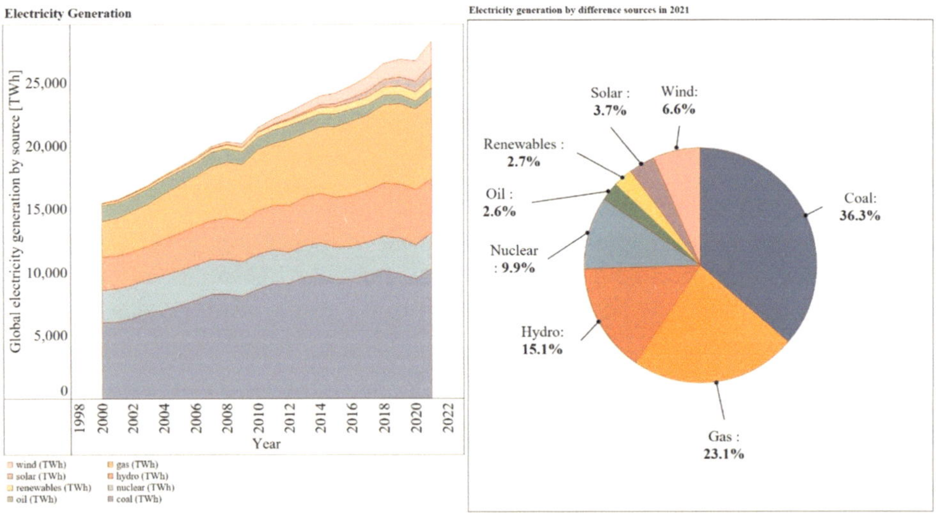

Figure 2.
Global electricity generation by sources from 2000 to 2021 [1].

of total primary energy sources. IEA reported that the increase in coal-fired power plants contributes to a sharp rise in carbon dioxide emissions. The electricity demand continues to grow by 4% in 2022. Despite substantial expansions of renewable energy usage, it is anticipated to offset the rise only partially in electricity consumption [2]. Due to the rise of greenhouse gas emissions, the environment is seriously threatened by the continued growth of energy consumption. Authorities from many countries, however, are focused on achieving net-zero carbon emissions and a major increase in the production of renewable and clean energy for end consumers. **Figure 3** shows that the total generated capacity will be 38,900 GW in 2050, while the expected rise in electricity output will be roughly 88,000 TWh.

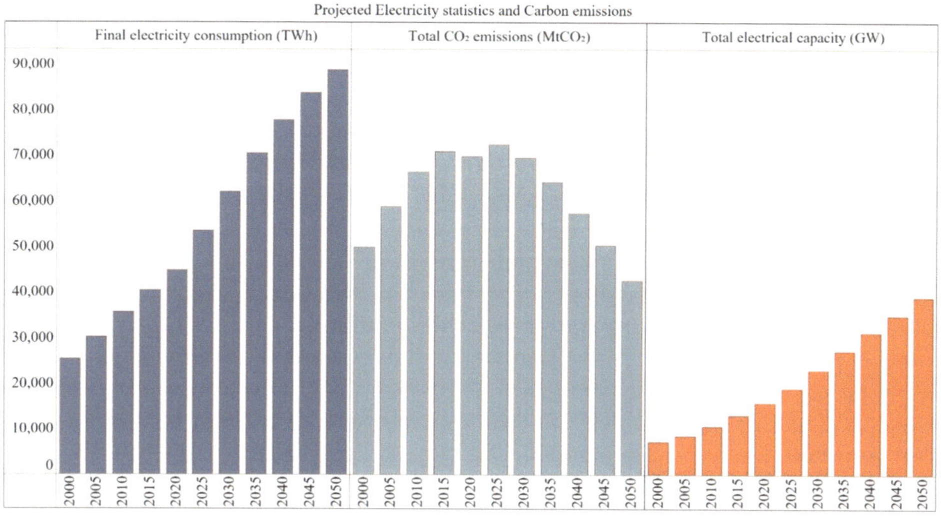

Figure 3.
Projected electricity statistics and carbon emissions till 2050 [3].

Additionally, it is anticipated that implementing the carbon tax will significantly reduce carbon emissions starting in 2025 [3]. In order to achieve net-zero carbon buildings, energy efficiency upgrades made to existing structures and energy-efficient designs for new buildings, including passive and active technology, will be crucial.

The built environment is seriously threatened by overpopulation and rapid urbanization. By 2050, the world's population is expected to reach 9.6 billion, a 21 percent increase from the current number. Therefore, the energy demand, particularly electricity for the built environment, will rise dramatically unless energy-saving options and measures are implemented. Moreover, 59% of the world population, as shown in **Figure 4**, resided in highly urbanized regions in 2020 because

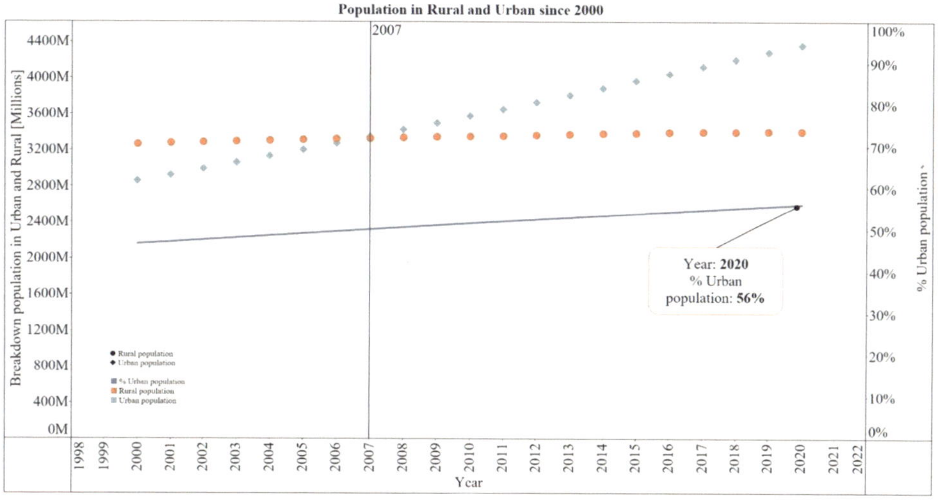

Figure 4.
World's Population residing in urban and rural areas [4].

these regions have employment opportunities, living standards, and ease of commute. After 2007, the proportion of urban residents overtook rural residents, sharply increasing the need for cooling and heating systems in residential and commercial structures. Urbanization significantly increased ambient temperature and decreased cooling system effectiveness due to the heat island effect. According to the IEA, two-thirds of homes may have air conditioning units [5]. By 2100, the average worldwide temperature could rise by 4°C due to the sharply increasing trend in the deployment of air conditioning systems in urban areas. Therefore, there is an urgent need to implement smart and energy-efficient air conditioning systems, including both passive and active cooling systems, for existing and new buildings. Doing so will lead to achieving net-zero carbon buildings.

Digitalization is a crucial component of the movement toward intelligent and energy-efficient solutions that are required to reach the targets of net-zero carbon emissions. Digitalization enables numerous energy systems to be more interconnected, intelligent, dependable, sustainable, and efficient. Digitalization could reduce energy consumption in buildings by around 10% by using real-time data to increase operational effectiveness. The installation of smart thermostats can also better predict heating and cooling requirements by employing self-learning algorithms, and real-time weather forecasts to predict occupant behavior.

2. Application of artificial intelligence in air conditioning systems

Machine learning (ML), a subset of *artificial intelligence*, widely applies to various sectors. The development of instrumentation and sensors has led to a significant increase in the amount of data collected per minute. Plotting and analyzing these data is crucial to turn them into insightful information that can be used for planning, operations, and forecasting. Machine learning techniques provide the link between the input parameters and the predicted output variables. Machine learning can be generally categorized into two groups, namely (i) supervi*se*d learning and (ii) unsupervi*se*d learning.

By deploying the appropriate methods, ML can be applied to the followings:

i. Detecting the sale trends

ii. Time series forecasting

iii. Multivariate time series forecasting with recurrent neural networks (NNs)

iv. Detecting financial fraud using decision trees

v. Convolutional neural networks implementation for car classification

Globally, many countries are embracing digitization, which will help businesses increase productivity, lower operating costs, and improve safety. Additionally, researchers and industry participants can create machine learning and artificial intelligence algorithms using historical data because it is easier to acquire thanks to digitization. However, even though machine learning is developing quickly, there have been difficulties using it in practical applications because it needs a vast amount of data. However, the road to digitization greatly aided the oil and gas industries' ability to access data, leading to machine learning easily. The built environment has been the focus of extensive research into intelligent control for air conditioning systems since 2000 to increase the effectiveness of these systems. Artificial intelligence applications in the HVAC sectors are made possible by digitalization, which is essential. Therefore, HVAC firms may create smarter systems to make buildings more environmentally friendly, thanks to technological improvements. Artificial neural networks (ANNs) have also been used in HVAC systems to optimize the operation set points of the air conditioning system.

2.1 Review of the application of machine learning (ML) and artificial intelligence (AI) in air conditioning systems

Many researchers have been working on machine learning and artificial intelligence for both the demand and supply side of HVAC systems. A vast majority of research conducted in the last 10 years can be generally categorized into (i) prediction of occupancy and their behavior, energy consumption, and energy management and (ii) control and optimization of HVAC systems.

Aftab et al. designed and implemented a sophisticated occupancy-predictive control system with the aid of recent development in embedded system technologies [6]. The system is cost-effective, has fewer requirements for powerful processors to execute highly sophisticated tasks, and deploys real-time occupancy recognition using video processing and ML techniques. The model can predict the occupancy pattern and allow to control of HVAC systems using real-time building thermal response simulations, achieving significant energy savings. Reeba et al. developed a model that can determine the occupants' behavior, which generally results in the wastage of energy in the operation of HVAC systems [7]. An ML-based model focused on the space's heat flow and could capture the energy waste depending on the status of the space, such as occupied or non-occupied. The model could predict the optimal temperature settings utilizing the status of the space, along with predicted mean vote (PMV) and the deployment of motion sensors. The author observed that about 50% of the total energy was wasted due to the suboptimal temperature settings in the space. Esrafilian-Najafabadi also analyzed the impact of different occupancy prediction models using ML techniques [8]. Four different ML techniques, namely decision trees, k-nearest neighbor (KNN), multilayer perceptron, and gated recurrent units, were deployed to predict the occupancy types and patterns and provide an accurate and reliable evaluation of the performance of the occupancy model for coupling with HVAC control systems. The author studied different models that analyze the occupants' energy savings and thermal comfort. The study included thermal comfort favored mode and energy savings priority mode. Despite having a trade-off between the occupants' energy savings and thermal comfort, the author observed that equally weighted energy savings and thermal comfort provide the best performance and that the KNN technique outperformed other machine learning techniques. Although numerous studies related to ML techniques that account for occupant patterns and behavior have been conducted, there is a lack of study on effective air distribution due to the dynamics of occupant patterns and their impact on temperature profiles across a spacious open office.

Many researchers emphasize their research on predicting energy consumption and optimizing energy usage by HVAC systems, the most energy-intensive system, utilizing supervised learning methods. For example, Liu et al. applied Deep Deterministic Policy Gradient (DDPG) for short-term energy consumption of HVAC systems [9]. The authors deployed a powerful autoencoder (AE) to process the raw data linked up with the DDPG method to attain high-level space state data for optimizing the prediction model. In this study, the authors set up a ground source heat pump system (GSHP) to supply a small office's cooling and heating needs. The operation data were used to train the model, and the authors demonstrated the office's energy consumption verification. The authors also verified that the proposed model predicted the state space variables more accurately than the common supervised learning models, such as support vector machine (SVM) and neural network (NN). The rapid expansion of deep learning techniques has made them promising alternatives to conventional data-driven methods. Vazquez-Canteli et al. developed an integrated simulation environment that links the building energy simulators and TensorFlow, which allows the implementation of various advanced machine learning algorithms [10]. This development enables many researchers to test and formulate optimized control algorithms to accommodate potential energy savings in buildings. The simulation platform also can be easily scaled up to the district or city level to study model-free algorithms and their impact on energy consumption and control strategy. Despite many interesting applications of ML and AI in HVAC systems being conducted, some research focused on energy consumption while other emphasized thermal comfort for small offices. There is a gap to close the loop between energy consumption while maintaining the thermal comfort, along with optimized cooling load predictions. In addition, most of the algorithms operate offline and cannot account for the heat loads in space's extremely dynamic nature and external parameters such as weather conditions. In order to incorporate artificial intelligence focused control that enables online load forecasting for extremely dynamic environments, this work is motivated by the desire to investigate the performance of HVAC systems, particularly airside systems.

3. Case study: deploying AI solution to airside energy efficiency improvement

3.1 Energy consumption in air conditioning systems

As illustrated in **Figure 3**, by 2050, worldwide power consumption is expected to have doubled from what it is today. Although it is questionable whether the sharp *rise* in power use is related to the sharp *rise* in cooling and heating requirements, the fact that there are currently 1.9 billion air conditioning units worldwide serves as proof. Additionally, it is anticipated that by 2050, cooling and heating requirements will have increased by 37%. Therefore, the *road map for* net-zero carbon buildings requires immediate effort to increase the efficiency of the air conditioning systems and the occupants' behav*ior*, incorporate cutting-edge control systems, and embrace passive technologies. This project aims to increase air conditioning system's efficiency by integrating them with AI-focused self-learning control systems.

A case study was carried out at one of the spacious offices at the Singapore Institute of *Technology (SIT)* to apply AI to air conditioning systems. Due to its tropical climate, space cooling is required throughout the year, and the building sector accounts for 37% of total energy consumption. **Figure 5** depicts the energy consumption of the air conditioning system, which is as high as 50% of building energy consumption due to its hot and humid climate. A detailed breakdown of the energy consumption of HVAC systems is shown in **Figure 5**, and airside accounted for 34% of the total energy consumption of HVAC systems. Although the airside energy consumption is *equall*y important compared to the waterside, it is mostly overlooked due to the high dynamics in nature.

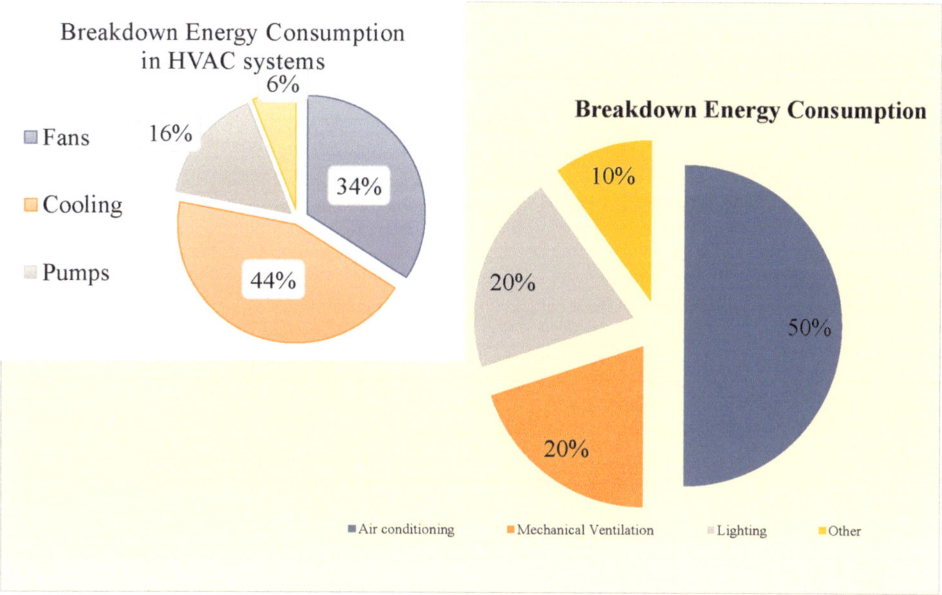

Figure 5.
Breakdown of energy consumption of buildings in Singapore [11].

3.2 Motivation of the case study

In chiller plants, airside systems account for the *second highest* energy consumption. In addition, the airside cannot support more control flexibility due to the high dynamics involved. Another thing to consider is that uneven thermal heat maps caused by oversized and undersized air distribution systems deviate from the thermal comfort of the occupants. **Figure 6** shows that hot and cold spots in large open offices are prevalent issues in airside systems.

Although there are other potential contributing elements, ineffective air distribution systems are the main problem. Ineffective air distribution systems cause hot areas because of insufficient cold air provided to the space. In addition, cold spots develop in the remaining areas of the office because there is excessive cold air. The control system, however, is unable to respond appropriately. Conventional control systems operate in

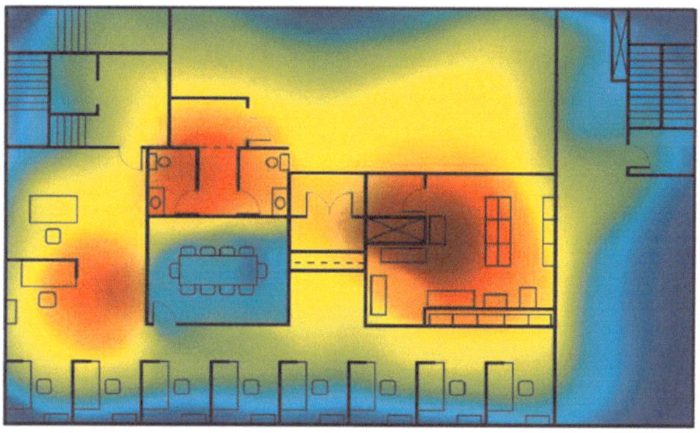

Figure 6.
Uneven thermal heat map due to improper sizing of AHUs and control strategies.

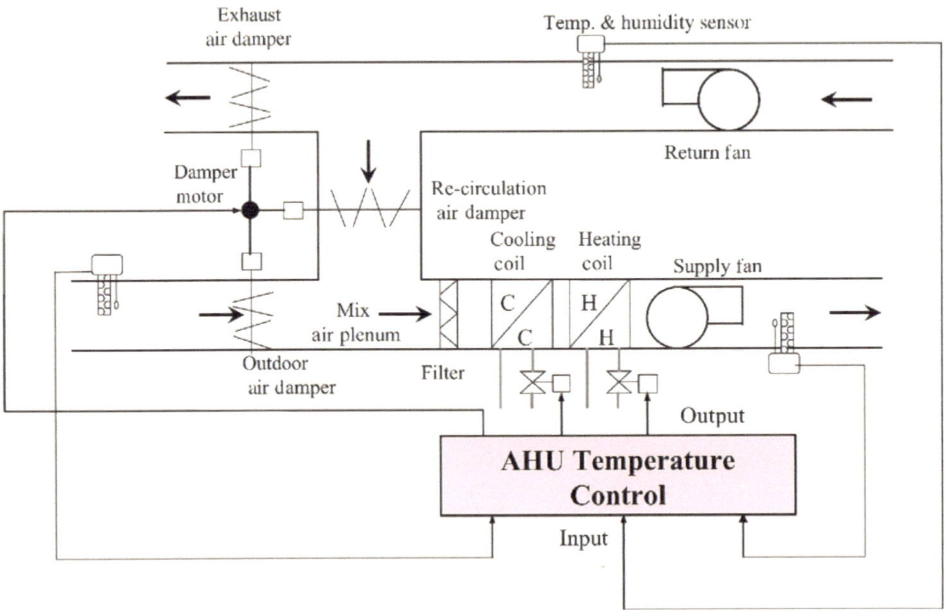

Figure 7.
Typical reactive feedback control system in air handling units (AHUs).

reactive methods, which is the cause of the control system's slow response. As indicated in **Figure 7**, although the main return air temperature is employed as feedback to serve as the control, it does not accurately represent the local zones, leading to unequal temperature distributions. The zoning of the space is one aspect that shows a significant role. The zone size is too large for the control systems to capture all information; thus, they are unable to adapt. Therefore, in this study, the impact of the big zone being segregated as smaller zones (*micro-zones*), proactive AI control on the performance of the airside system, and energy savings potential will be investigated.

3.3 Details of the testbed located at the Singapore institute of technology

The pilot tests were conducted on one floor of the Singapore Institute of Technology (SIT) campus located at *Dover Drive*. The *testbed occupying* 11,000 square feet is located at *level two*, compris*ing* open offices, meeting rooms, a pantry, an AHU room, and washrooms. The space is fully air-conditioned except for the washrooms. The details of the testbed are tabulated in **Table 1**. **Figure 8** illustrates the layout of the

Total floor area	**11,000 square feet (sqft)**
Seating capacity	Approx. 100
Operation hours	0830 hrs to 1800 hrs
Area types	Enclosed workstations, cabins, cafeteria, and conference rooms
BMS	Yes, Johnson controls
Chilled water actuator	Yes, installed for each AHU

Table 1.
Details of the pilot in the case study.

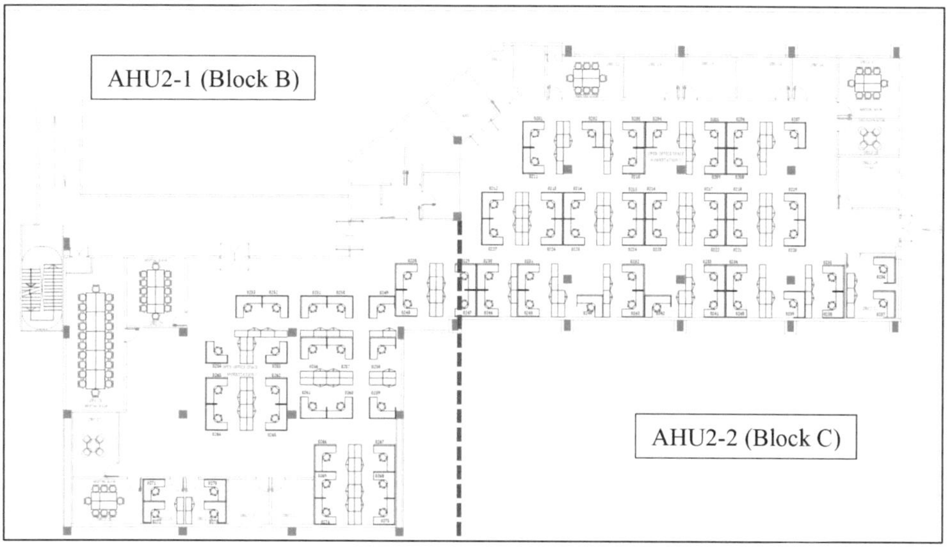

Figure 8.
Plan view of the office space at level two of the University Service Centre at the Singapore Institute of Technology.

office located at the Singapore Institute of Technology, and the spaces are segregated into two zones, namely, *block B and block* C. While block B's cooling requirements are supplied by air handling unit 2-1 (AHU 2-1), block C is served by AHU 2-2. The temperature set point of the space was 24°C throughout the day.

The key issue with the air conditioning system is the thermal comfort of the occupants stationed in the space. From the occupant's feedback, it is discovered that there are areas with hotspots and overcooling within the office. On occasion, occupants feel uncomfortably hot or extremely cold in the office. It is observed that some of the diffusers are covered with masking tape to restrict airflow. The AHU VFD and actuator set points are changed manually based on complaints from the occupants. In addition, due to the work nature of the academic staff, they are frequently required to leave and return to their desks for lectures and classes, resulting in a dynamic heat map. Therefore, there is a need to resolve the issue without compromising the energy efficiency of the air conditioning system. The primary objective of this study is to develop an intelligent solution to resolve thermal comfort issues without compromising energy efficiency while eliminating the conventional reactive approach to control systems.

The proactive solution would account for the varying occupant numbers throughout the day while creating an optimal condition for their staff. Despite the abundant availability of smart sensors, which work on room levels, an AI algorithm was developed and tested at SIT staff office, along with the collaboration between SIT and Singapore Digital Pte. Ltd., a sole distributor of 75F smart innovation solutions in Singapore.

Dynamic air balancing and chilled water balancing, along with proactive AI predictive control, are the essential components of this study to achieve energy savings while maintaining the thermal comfort of the occupants. In order to optimize the air distribution efficiency, two big spaces, as indicated in **Figure 8**, are divided into 43 micro-zones, as indicated in **Figure 9**. Each meeting room is treated as a micro-zone. Each micro-zone in the open office is equipped with an IoT smart sensor that measures the key parameters, such as temperature, relative humidity, CO_2 concentration, and occupancy status, using a passive infrared sensor (PIR), enabling an accurate representation of the local heat load. Moreover, as illustrated in **Figure 10**, smart

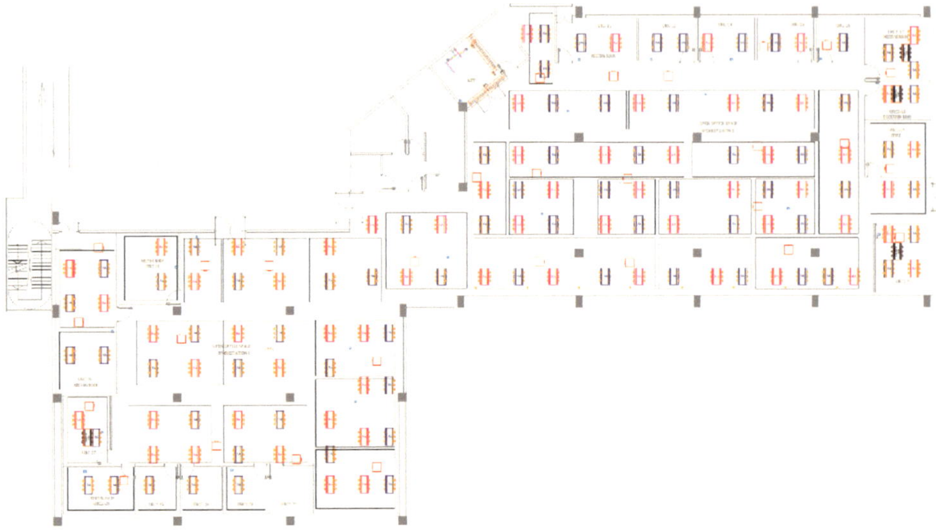

Figure 9.
Two zones are split into 43 micro-zones.

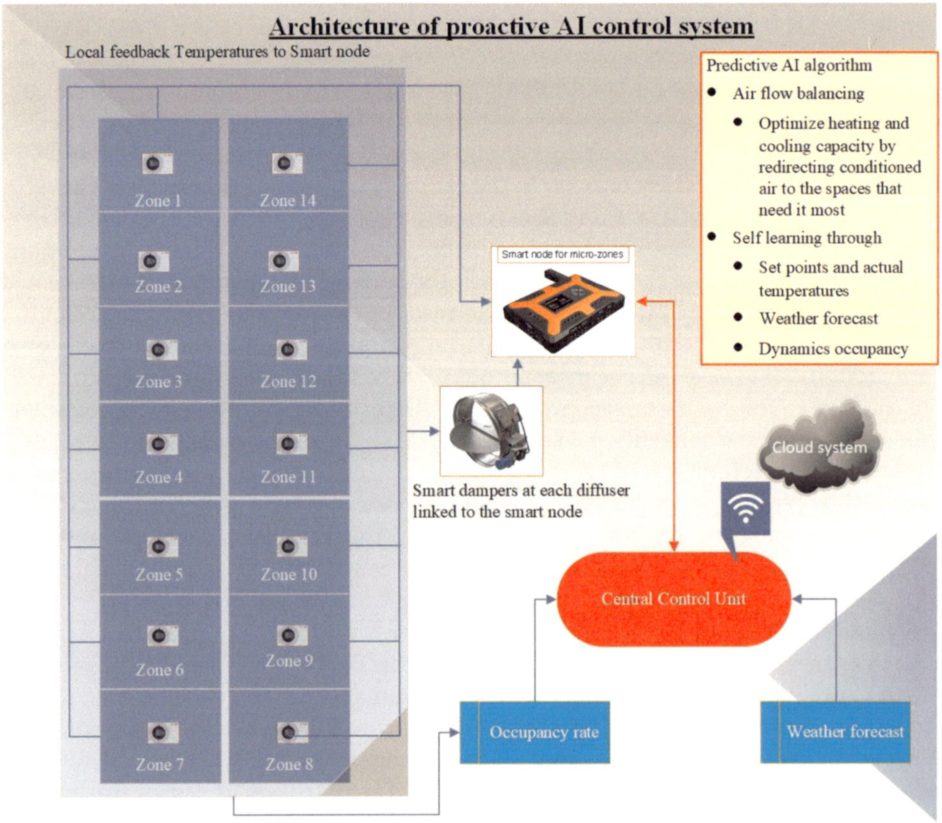

Figure 10.
Architecture of proactive AI control system.

dampers are retrofitted between each supply air diffuser/VAV duct and flexible air duct to modulate the amount of airflow based on the actual heat load. This facilitates micro-zonal control, allowing better comfort and energy savings. In addition, the opening of the smart damper is controlled based on the local heat load. The IoT smart sensors and dampers are wirelessly connected smart nodes which communicate wirelessly with central control units (CCUs). A cloud-based proactive AI control powers the algorithm behind the control units, and the architecture of the proactive AI control system is shown in **Figure 10**. The CCU sends minute-by-minute data regarding temperatures in various building parts to cloud servers. Every night, these servers run proprietary algorithms to crunch the historical data and develop a thermal model of the building. They then predict the thermal load in each part of the building for the next day based on the forecasted weather.

3.3.1 Dynamic airflow balancing

Figure 11 illustrates the dynamic airflow balancing, which optimizes the cold air supply to the most required space. The smart dampers' opening at micro-zones with cold spots is adjusted to accommodate the cooling needs in that micro-zone. Due to the changes in the opening of the smart dampers, the static pressure in the duct increases. However, cold air is circulated to space (hotspots), which requires more cooling, restoring the static pressure. Therefore, supply air fan speed is not ramped up to supply more cooling to the hot space; instead, air balancing between cold and hotspots progresses, resulting in energy savings in AHUs. This means that the AI control is able to identify which zones require more cooling by deploying dynamics zone *priority* (DP). Since the system enables minute-*by-minute data* collection, real-time DP is performed prior to executing the next control phase. Air balance is performed using a weighted average of the local heat load, as shown in Eq. (1) (**Figure 12**).

$$\dot{Q}_{weighted} = \frac{\sum_{i=z_1}^{z_n}\left(\dot{Q}_{i,j}xDP_i - \dot{Q}_{i,k}xDP_i\right)}{\sum_{i=z_1}^{z_n}DP_i} \quad (1)$$

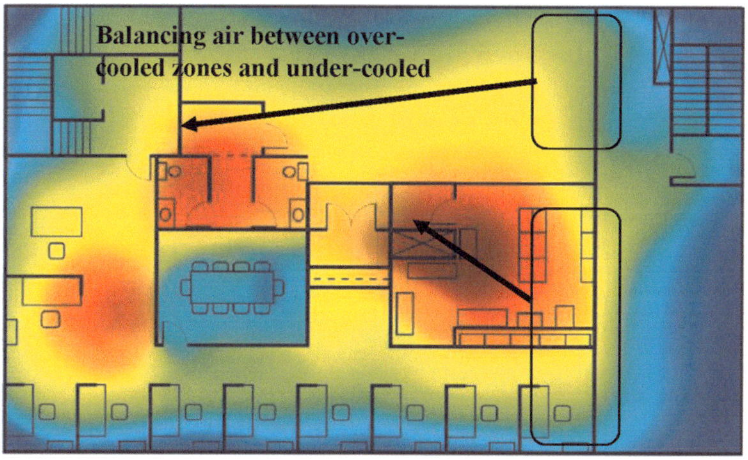

Figure 11.
Illustration of dynamics of airflow balancing.

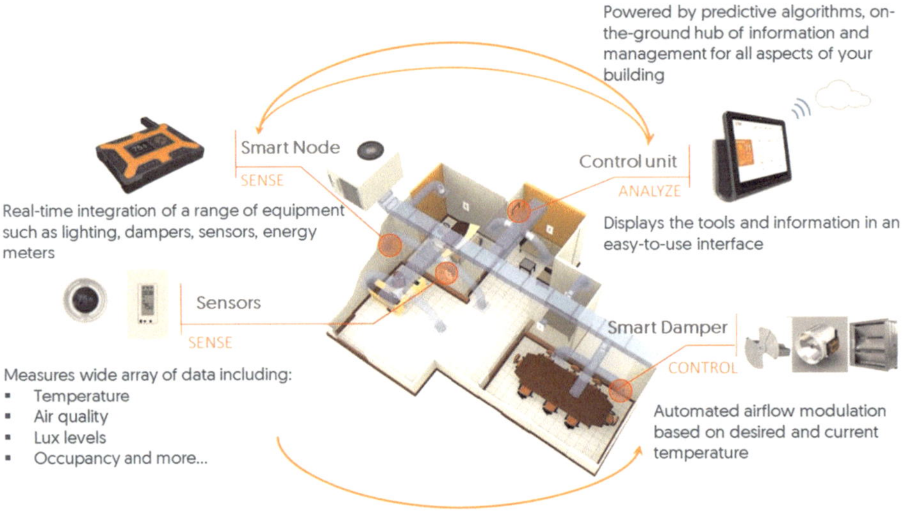

Figure 12.
Dynamic airflow balancing diagram.

where $\dot{Q}_{weighted}$ and $\dot{Q}$ denote the weighted average of heat load and local heat load in the space, respectively, DP refers to dynamics zone priority, i denote the number of zones in the space, j represents the zones with overcooling, and k is for the zones requiring more cooling. Then, air balancing for the micro-zones is carried out based on the DP value, which identifies how far the current temperature is away from the set point. The AI algorithm identifies and optimizes the air balancing, resulting in the evenly distributed cold air supply to each of the micro-zones, and AHUs can still be operated at a lower speed as compared to the conventional control system because the speed of the AHUs is adjusted, as shown in **Figure 13**, based on the weighted average of micro-zones after air balancing is carried out. In addition, fresh air optimization is enabled by incorporating a modulating damper in the fresh air duct. The bandwidth of the opening of the fresh air damper ranges from 20 to 100% based on CO_2 concentration in the space, enabling minimal fresh air usage when the indoor CO_2 level is about 900 ppm.

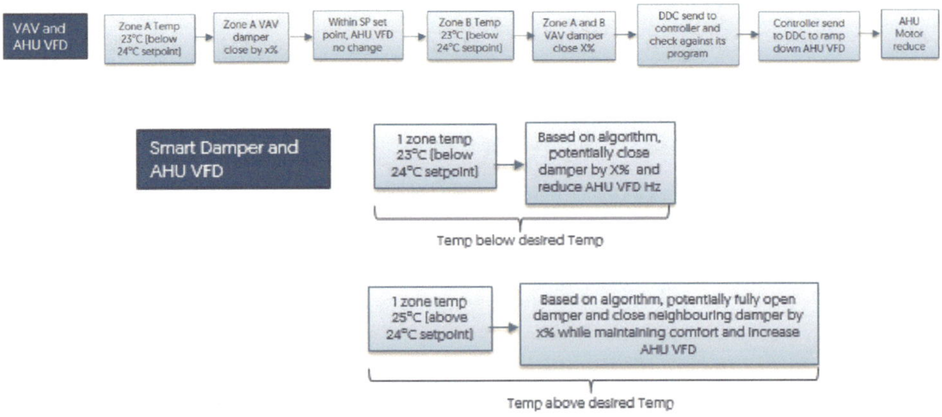

Figure 13.
Block diagram that shows smart damper and AHU VFD relational control.

3.3.2 Dynamic chilled water balancing

Chilled water balancing is achieved by utilizing the micro-zones' weighted average return air temperature. When the weighted return air temperature falls above the set point, the Al algorithm detects that more cooling is required in the space. However, the steps for air balancing and AHUs speed adjustment are completed to accommodate the cooling requirement. Therefore, the controlled valve will be modulated to a wider position to provide more chilled water to maintain the temperature in micro-zones within the thermal comfort range defined by ASHRAE standard 55 [12]. Therefore, the differential temperature of the chilled water is maintained at the optimal range, while the chiller water pump's speed is adjusted to provide the required cooling in the space, resulting in energy savings without compromi*si*ng the thermal comfort of the occupants in the space. The sequence of activating the opening of the chilled water modulating valve is shown in **Figure 14**.

3.4 Deployment of AI solution in airside system of the chiller plant located at SIT

Implementation of measurement of the performance of AI-oriented proactive control solution comprises the following stages:

1. Retrofitting of smart dampers to the existing system that enables dynamics air balancing

2. Installation of a power meter at each AHU to measure the power consumption of the fans

3. Installation of CCU for each AHU to control smart nodes and smart dampers, modulate VFD and chilled water actuator, and act as a cloud gateway

4. Installation of smart dampers at the existing mixing boxes outlet. The opening and closing of the smart dampers are controlled by the smart nodes installed above the false ceiling. The smart nodes communicate wirelessly to the cloud; users can access all data and control through the App or portal.

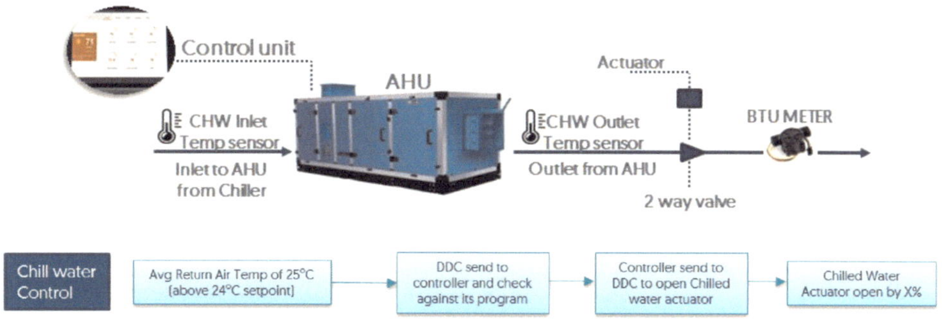

Figure 14.
Dynamic chilled water balancing control process.

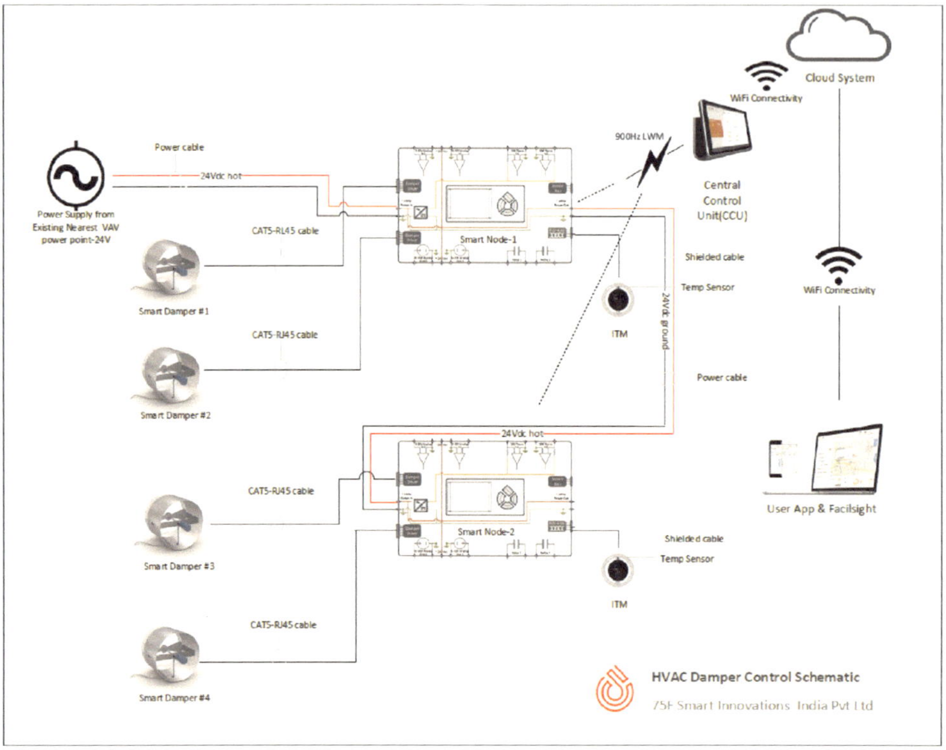

Figure 15.
A schematic diagram of dynamic airflow balancing.

5. Installation of Intelligent Temperature Mote (ITM) for each zone across the pilot area to sense and collect data (temperature, humidity, and Lux) in real-time every minute, and there are a total of 43 zones, as shown in **Figure 15**.

6. Installation of chilled water flow meter, the temperature sensors for chilled water return, and supply.

7. Installation of a chilled water actuator controlled by the CCU to modulate chilled water flow to match the optimal set point and ensure optimal flow rate and differential temperature through the chilled water pipe networks, as shown in **Figure 16**.

8. Installation of a new fresh air damper with Belimo actuator and the fresh air damper is modulated (20–100% opening) based on the CO_2 level, as shown in **Figure 17**.

3.4.1 Baseline measurement and smart mode measurement deployment of AI solution in the airside system of the chiller plant located at SIT

After completing the installation of the required instrumentation, sensors, and IoT devices and commissioning, which includes fine-tuning the parameters, the testbed was operated in two phases: baseline mode and smart mode. Each mode was operated for 10 days, excluding weekends. The baseline mode represents the operation of

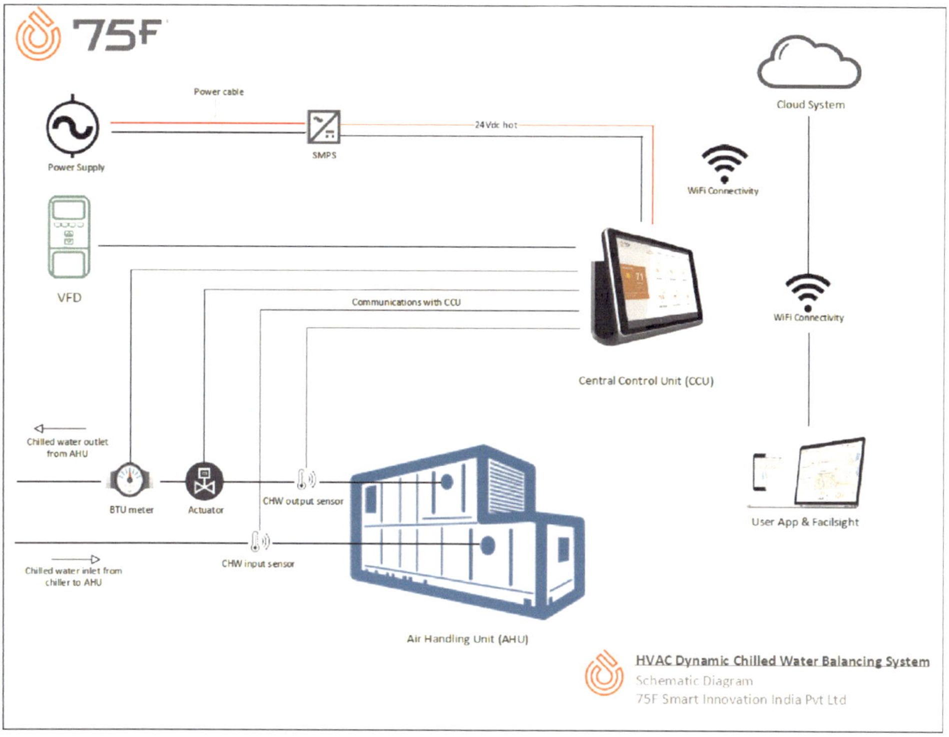

Figure 16.
A schematic diagram of dynamic chilled water balancing.

existing conditions, isolating the proactive AI control, whereas the smart mode enables the proactive AI control, including dynamic air balancing, dynamic chilled water balancing, and fresh air optimization. The AI control overwrites the set point of BMS for existing operating conditions. Outdoor temperature and relative humidity were also recorded in the cloud during the testing of both phases to ensure that the impact of the weather conditions on the airside system's performance was considered. During both testing phases, data are recorded every minute using the instrumentation and sensor installed during the retrofitting stage, as tabulated in **Table 2**. During weekdays, AHUs are scheduled to start at 6:30 am, and the chiller and pumps are staged to turn on from 7 progressively for pre-conditioning. Since the building operates from 8:30 am to 6:00 pm, the data analysis only includes this period of the day. Key parameters, such as the temperature and relative humidity of all 43 zones, were recorded every minute in both baseline and smart modes. While AHU 2-1 supplies the cooling requirements to zone 1–23, AHU2-2 serves zones 24–43. The set point for all spaces was maintained during the tests at 24°C.

Figure 18 depicts the temperature profiles of supply air measured during the baseline and smart mode tests. It is indicated that supply air temperature fluctuated between 13.6°C and 21.7°C, whereas it was maintained between 15.3°C and 19.3°C. The median temperature of supply air for baseline test and smart modes were 16.3°C and 17.1°C, respectively. Despite maintaining the close median supply air temperature between baseline mode and smart mode, the differential temperature in the interquartile for baseline mode was 1.5°C, and that of smart mode was about 1°C. It is also concluded from the box plot that most of the supply air temperature during the

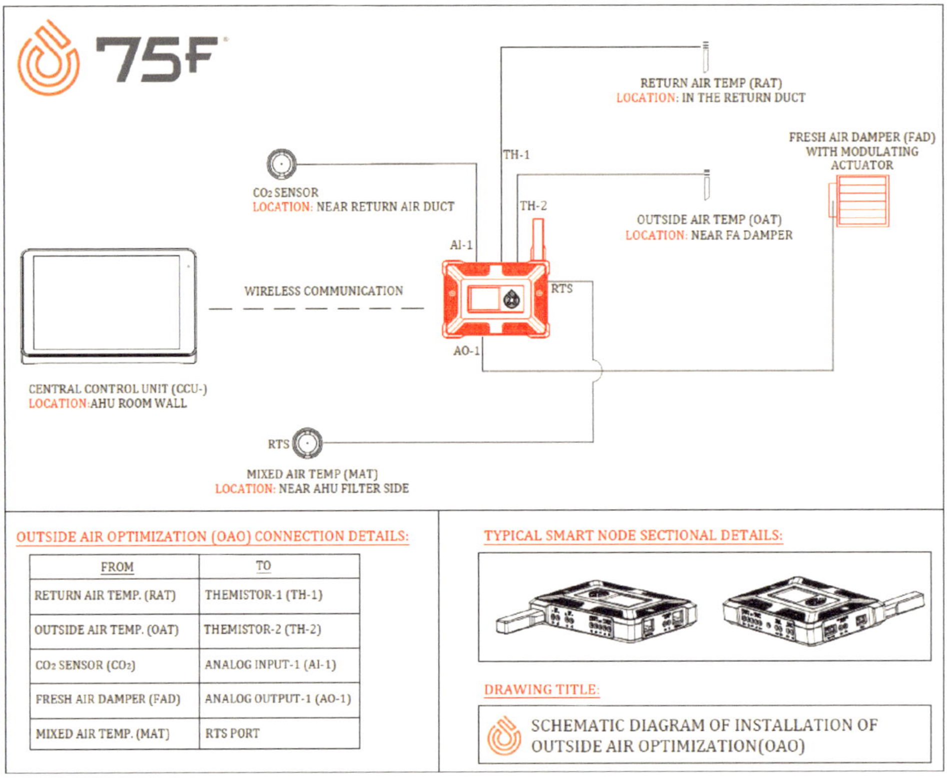

Figure 17.
A schematic diagram of uutside air optimization comfort range.

Location	Parameters measured
Meeting rooms and office space	• Temperature • Relative humidity • CO_2 concentration • Dynamics occupancy • Supply air temperature from each diffuser • Smart damper opening in %
AHUs	• Fan power • Fan speed • Chilled water flow rate • Chilled water supply and return temperature • Supply air temperature • Return air temperature • Fresh air damper opening in % (20–100%)

Table 2.
List of measured parameters and locations of measurement.

smart mode test fall outside the interquartile, and outliers are beyond 1.5 times the interquartile (upper whisker) due to inefficient air distribution systems and control strategy.

On the other hand, no outliers were discovered beyond the upper and lower whiskers during the smart mode testing. It is also worth noting that the temperature

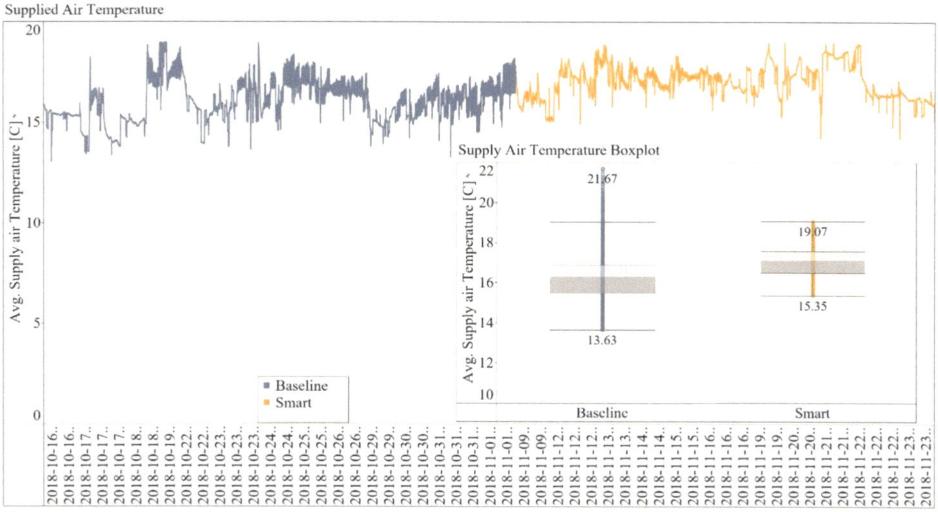

Figure 18.
Supply air temperature profiles during the baseline and smart test.

difference between the minimum and maximum supply air temperature was less than 4°C, assuring that smart mode control performs significantly better than baseline mode in terms of air distribution effectiveness. The space temperature with respect to time during baseline mode and smart mode is presented in **Figure 19**. During the test period of both modes, the set point temperature was maintained at 24°C, and the results were analyzed by comparing the baseline and smart mode tests. From the temperature and relative humidity profiles during the baseline test, it was observed that the space temperature during the smart mode test fluctuated from 21 to 25°C, while relative humidity in the space varied between 67% and 48%. Furthermore, the difference between space temperature and the set point was found to be considerably huge in some cases; it was as high as 3°C, resulting in cold spots and hotspots in space. However,

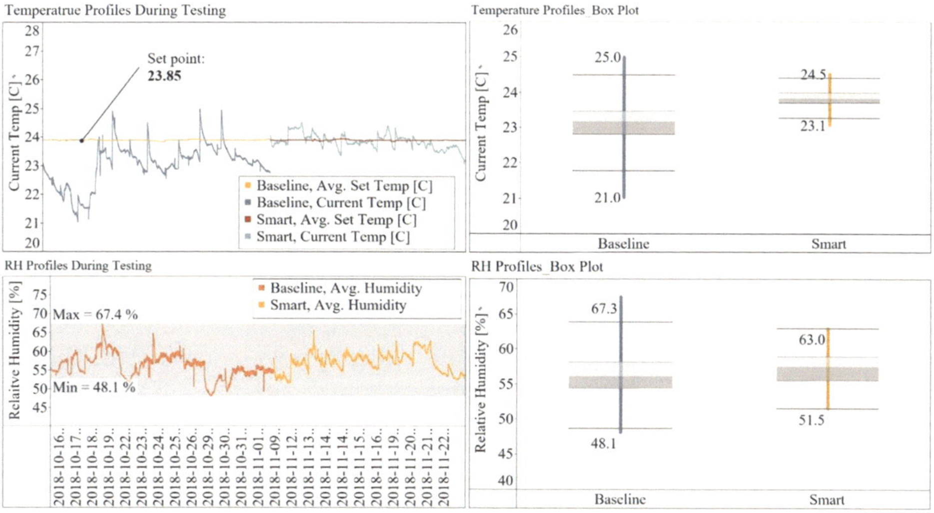

Figure 19.
Average temperature and relative humidity profiles of the space during the baseline and smart mode tests.

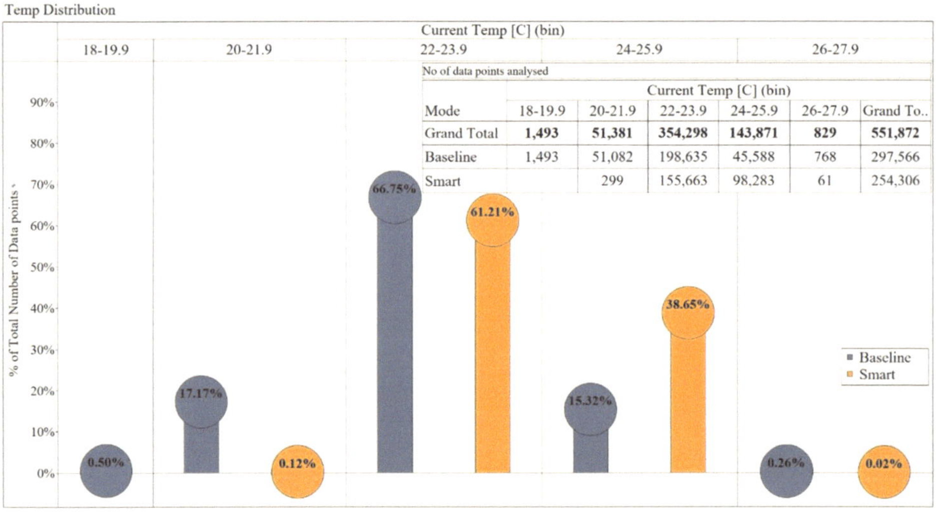

Figure 20.
Temperature distributions with 2°C range.

during the smart mode test, the space temperature varied between 23 and 24.5°C, while relative humidity ranged from 52 to 65%, which falls well within the thermal.

In order to analyze further details of the temperature distribution in the space, a temperature bin is created with 2°C range with a total of 551,872 data points, and the results are illustrated in **Figure 20**. Seventeen percent of the data points that falls under undercooled regions (19–22°C) during the baseline test were shifted to 22–26°C when the smart mode was activated. Moreover, the smart mode delivered 99.97 percent of the events within the bin range of 20–23.9°C, highlighting that proactive AI control works perfectly fine to optimize the airside performance compared to the baseline mode. Therefore, proactive AI control not only achieves a better thermal comfort condition in the space but also improves the efficiency of the airside system, because AI control optimizes the cooling load prediction by adapting the characteristics and activity ongoing in the space, along with the dynamic airflow balancing strategy. Energy consumption should not be overlooked despite improving the thermal performance of the airside. Therefore, energy data, such as electricity consumption and cooling supplied to the building, were monitored and recorded throughout both baseline and smart modes. All energy data were recorded using the Kamstrup BTU (cooling energy) and the Schneider Energy Meter (electrical energy). Data during the weekends of the testing period were excluded from the analysis in both modes. Two AHUs (AHU 2-1 and AHU 2-2) were assigned to supply cooling to the space, and the rated power of AHU 2-1 and AHU 2-2 at the full load are 5.7 kW and 3.7 kW, respectively. During different test modes, weather conditions were normalized to ensure that the deviation in the weather conditions was not affected. The pairs of the daily average ambient temperatures during both modes for comparative analysis are presented in **Figure 21**.

Daily electricity consumption of both AHU 2-1 and AHU 2-2 is illustrated in **Figure 22**. During the baseline test, it is observed that the daily electricity consumption of AHU 2-2 ranges between 44 kWh and 70.50 kWh, while the electricity consumption of AHU 2-1 varies between 22.6 kWh and 8 kWh. While conducting the test in the smart mode, as indicated in **Figure 22**, electricity consumption of AHU 2-2

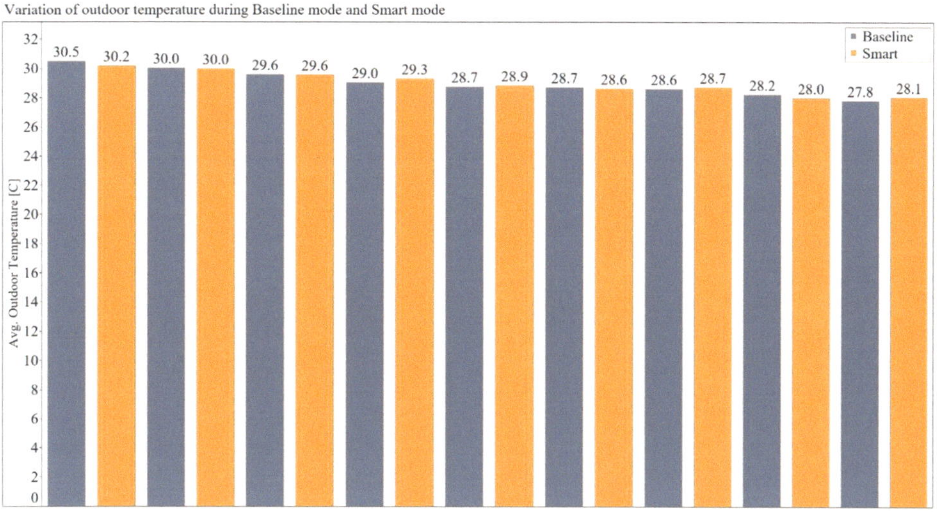

Figure 21.
The pairs of average daily outdoor temperatures for the comparative analysis during the baseline and smart modes.

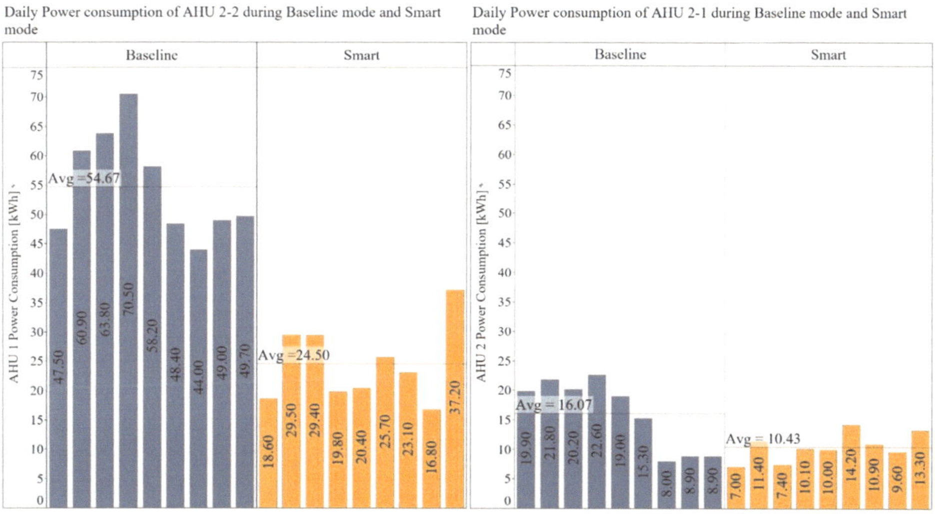

Figure 22.
Daily electricity consumption of AHU2-1 and AHU2-2 during the baseline and smart modes.

fluctuates between 37.2 kWh and 18.6 kW and that of AHU 2-1 peaks at 14.2 kWh, and its minimum value is 7 kWh. The average electricity consumption of AHU2-1 and AHU 2-1 during the baseline mode was 54.67 kWh and 16.07 kWh, respectively. However, the average electricity consumption of both AHUs during the smart mode was 24.5 kWh [AHU2-1] and 10.43 kWh [AHU 2-2]. Therefore, the total electricity consumption of AHU 2-2 is cut from 492 kWh in the baseline test to 220.5 kWh in the smart test, whereas the electricity consumption of AHU 2-1 is lowered by 50.7kWh from 144.6 kWh to 93.9 kWh, as demonstrated in **Figure 23**. The results also highlight that electrical energy savings in AHU 2-2 are about 55%, while AHU 2-1 saves approximately 35% of electricity usage when the smart mode is activated. While

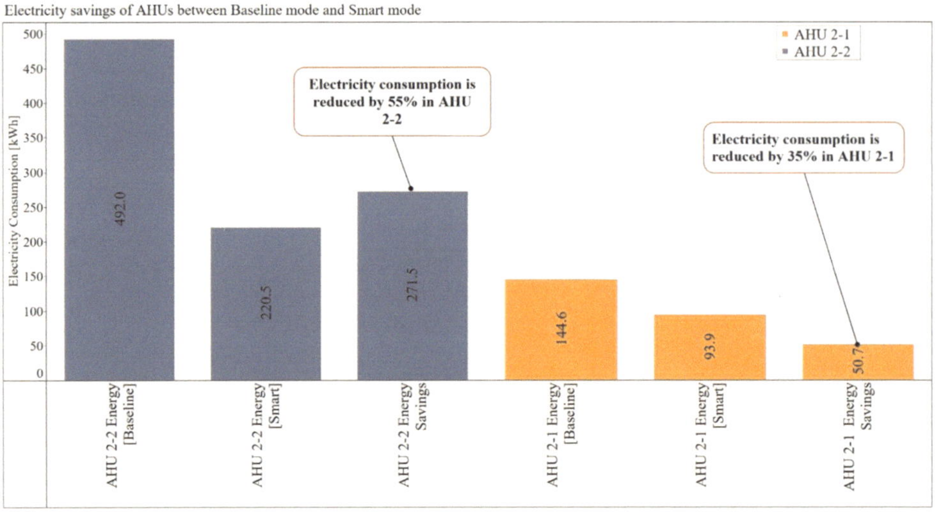

Figure 23.
Electrical energy savings at different AHUs between baseline and smart modes.

presenting electricity consumption analysis, cooling energy consumption is also investigated in this case study. Due to some constraints in the installation of BTU meters for each AHU to measure the cooling energy, only one BTU meter was installed at the common chilled water header to log the chilled water flow rates. Therefore, the cooling energy consumption (kWh) is calculated as follows:

$$\dot{Q}_{cooling} = \dot{m}_{chw} Cp_{chw} (T_R - T_S) x N_{op} \tag{2}$$

In Eq. (2), the first parameter $\dot{Q}_{cooling}$ represents cooling energy consumption in kWh; the second parameter $\dot{m}_{chw}$ denotes mass flow rates of chilled water in kg/s, the third parameter Cp_{chw} is the specific heat capacity of chilled water in kJ/kg·K, T represents temperature in °C, and N_{op} is the operation time in hours. The subscript R and S represent return and supply, respectively. **Figure 24** illustrates the accumulative cooling energy consumption for the baseline and smart mode tests. The smart mode is observed to consume 29% less cooling than the baseline test while maintaining thermal comfort in the space, because the cooling requirements in the office are significantly reduced by optimizing the supply airflow rates to facilitate the cooling load in each micro-zone. The results show that airside energy consumption can be reduced by as high as 50% of electricity consumption in AHUs, while the reduction in cooling supply to the office was also approximately 29%. The results also assure that reduction in the cooling supply and electrical energy consumption do not compromise the thermal comfort of the office.

This case study demonstrated the application of AI-oriented control in airside air conditioning systems to resolve typical issues, such as thermal comfort and high energy consumption due to overcooling and undercooling, in open offices. It also highlights that the improvement on the airside also contributes to the reduction of electricity consumption of the fans, resulting in minimizing the waste energy as compared to the baseline control system while cooling required in the offices is also optimized.

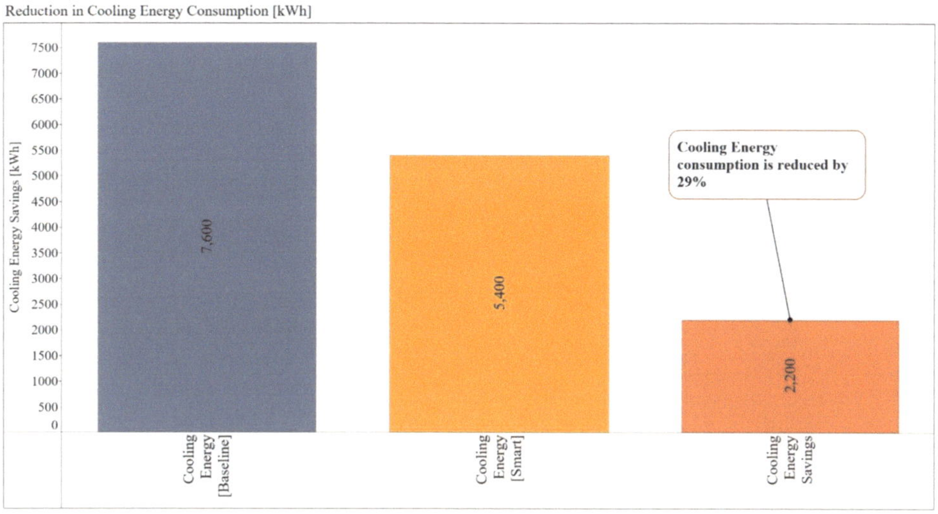

Figure 24.
Cooling energy consumption during baseline test and smart test.

4. Conclusion

This chapter investigates the application of AI solutions in optimizing airside performance while maintaining thermal comfort in the office. Pilot tests were conducted to examine the impact of proactive AI control on resolving common thermal comfort issues in the office, such as overcooling and undercooling. The demonstration testbed was implemented in one of the Singapore Institute of Technology floors, located at Dover, Singapore. The tests were conducted in baseline mode (conventional BMS control) and smart mode (proactive AI control). The results highlight that the proactive AI control solution provides not only the improvement of energy consumption but also an enhancement in thermal comfort by eliminating cold spots and hotspots in the office. Furthermore, it also highlights that the improvement on the airside also contributes to the reduction of electricity consumption of the fans, resulting in minimizing the waste energy as compared to the baseline control system, while cooling required in the offices is also optimized.

Acknowledgements

The author would like to express his gratitude to SP group Pte. Ltd., SP Digital Pte. Ltd. and 75F solution for providing the financial support to accomplish such a live testbed on SIT campus.

Author details

Aung Myat
Singapore Institute of Technology, Singapore, Singapore

*Address all correspondence to: aung.myat@singaporetech.edu.sg

References

[1] Available from: https://ourworldindata.org/energy-key-charts

[2] Electricity Market Report – July 2021, IEA

[3] World Energy & Climate Statistics – Yearbook. 2022. Available from: https://yearbook.enerdata.net/

[4] Hannah Ritchie and Max Roser (2018)—Urbanization. Available from: https://ourworldindata.org/urbanization

[5] The Future of Cooling: Opportunities for Energy-efficient Air Conditioning, 2018

[6] Aftab M, Chen C, Chau C-K, Rahwan T. Automatic HVAC control with real-time occupancy recognition and simulation-guided model predictive control in low-cost embeddedsystem. Energy and Buildings. 2017;**154**:141-156. DOI: 10.1016/j.enbuild.2017.07.077

[7] Reeba R, Naveed ULH, Chau Y. Determination of consumer behaviour based energy wastage using IoT and machine learning. Energy and Buildings. 2020;**220**. DOI: 10.1016/J.ENBUILD.2020.110060. Available from: https://www.sciencedirect.com/journal/energy-and-buildings/vol/220/suppl/C

[8] Esrafilian-Najafabadi M. Haghighat F Impact of occupancy prediction models on building HVAC control system performance: Application of machine learning techniques. Energy and Buildings. 2022;**257**:111808. DOI: 10.1016/j.enbuild.2021.111808

[9] Liu T, Xu CL, Guo Y, Chen H. A novel deep reinforcement learning based methodology for short-term HVAC system energy consumption prediction. International Journal of Refrigeration. 2019;**107**:39-51. DOI: 10.1016/j.ijrefrig.2019.07.018

[10] Vázquez-Canteli J, Ulyanin S, Kampf J, Nagy Z. Fusing Tensor Flow with building energy simulation for intelligent energy management in smart cities. Sustainable Cities and Society. 2019;**45**:243-257. DOI: 10.1016/j.scs.2018.11.021

[11] Singapore's Second National Communication: Under the United Nations Framework Convention on Climate Change. National Environment Agency; 2010

[12] Thermal Environmental Conditions for Human Occupancy. ASHRAE Standard; 2020. p. 55

Section 2

Heating, Cooling and Air Conditioning

Chapter 3

Heat Recovery from Cryptocurrency Mining by Liquid Cooling Technology

Nan Chen, Yunshui Chen and He Zhao

Abstract

Bitcoin, the world's largest cryptocurrency, currently consumes an estimated 150 terawatt-hours of electricity annually. Most cryptocurrency miners have dissipated the thermal energy from mining chips to the ambient by air cooling circulation. To recover the thermal energy from cryptocurrency mining, an advanced heat recovery system has been developed, prototyped, and tested. The cryptocurrency miners in an enclosure are cooled by spraying dielectric coolant, then the coolant heated by the mining chips is collected and driven through the spiral heating coil immersed in a 190 L hot water tank. High efficient liquid spray cooling mechanism is the core of this design, by which maximum coolant temperature can reach 70°C in the field trail within the safe temperature limits of mining chips. In practice, this record temperature not only meets the minimum legionellosis risk management requirements for building water systems defined by ANSI/ASHRAE Standard 188-2018 but also provides high-grade energy input to the building, district heating system, or booster heat pump/boiler if needed. In theory, the conventional concept of PUE based on energy has been redefined by the PUE based on exergy. The energy-based PUE is 1.03 and the exergy-based PUE is 0.95 in this case, which can truly reflect the useful energy flow, exergy, in the heat reclaim system.

Keywords: heat reclaim, cryptocurrency miner, exergy, liquid spray cooling, PUE

1. Introduction

Bitcoin, the world's largest cryptocurrency, currently consumes an estimated 150 terawatt-hours of electricity annually-more than the entire country of Argentina, with a population of 45 million [1]. Acknowledging that Bitcoin mining is a high energy consuming and high energy density industry, correspondingly, miners generated large quantities of heat during the continuous hashing process, which conventionally has been dissipated into the ambient by air cooling circulation. Regarding the financial and environmental benefits from the heat reuse in mining industry, competitive entities are explored an economical approach to elevate the revenue performance and cut the carbon footprint at initial stage of project plan and make effort to maximize these benefits further by advanced heat exchanging technologies, which can partially or completely replace the traditional fossil fuels in the field of low temperature heating. Due to the resilience of spatial arrangement, miners can be deployed in the pattern of centralized and decentralized models. Centralized mining farms are more suitable to heat the green house, vertical farm, district heating network, i.e., work as a heat resource in a large scale. On the contrast, decentralized mining utilities can provide the space heating or hot water in the small-scale projects, such as in residential or light commercial deployment.

Heat recovery is a low-carbon technology that cuts across multiple sectors of the technologies. From the perspective of the first law of thermodynamics, the primary challenge is to identify effective ways to capture and transfer heat, along with sound economic use cases. From the perspective of the second law of thermodynamics, the second challenge is to elevate the grade of heat expelled by miners to extend the scope of waste heat application and provide feasibility to seamless connection with the current heat system driven by electricity or fossil fuels. Regarding the approaches of capture and transferring heat from miners, current technologies can be categorized into three types, i.e. air cooling, liquid immersion cooling (single phase and two phases), and hybrid (air/liquid) cooling [2]. Air cooling utilizes fans to pass air over the miner heat sinks. Hampus [3] reports that 5.5–30.5% of the electrical input to a 1 MW air-cooled mining farm could be recovered, which can fill the heating demand of a 2000m^2 greenhouse by 89.7–97.9%, and a 10,000 m^2 greenhouse by 50.0–61.5% respectively. Up to 94.5% of thermal energy has been wasted due to the poor thermal properties of air in the process of thermal transmission. Enachescu [4] pointed out that waste heat from a 45 MW data centre is sufficient to provide a year-round heating to a 8.34-acre greenhouse for commercial cannabis growth. The total annual avoided emissions were calculated at 70,000 tonnes of CO_2 by economizer cycles and waste heat reuse in Alberta, Canada. Agrodome [5] is the first retail facility that uses servers as heat sources distributing residual heat through a cascading set of greenhouse applications. Blockchain Dome, each dome has an input of 1.5 MW and produces 5,000,000 BTU/h of heated air, requires no additional electricity to maintain the preferred temperature range. In July of 2018, United American Corp. [6] announced their intention to deploy 25 Blockchain Domes across Quebec, filing a power license request at 5 MW at the large power preferential rate. Data center heat reuse co-location with an associated greenhouse minimizes losses to electricity distribution, heat transportation, and associated heat loss, which is a primary scenario for heat recovery from the air-cooled data center.

Immersion cooling is an IT cooling practice by which IT components and other electronics, including complete servers and storage devices, are submerged in a thermally conductive but electrically insulating dielectric liquid or coolant (single-phase or two-phase). Heat is removed from the system by circulating relatively cold liquid into direct contact with hot components, then circulating the now heated liquid through cool heat exchangers. The advantages of using liquid cooling over air cooling include liquid's higher specific heat capacity, density, and thermal conductivity. This allows liquid to transmit heat over greater distances with much less volumetric flow and reduced temperature difference. Regarding the higher coolant temperature, inlet water temperatures [7] in the order of 45–70°C can cool server rack chips and CPUs to approximately 80–90°C to avoid triggering the auto frequency throttling protection, while the maximum permissible temperature range of processors is 100–120°C. According to [8], the inlet temperatures can range between 30 and 60°C. The solution presented by Ernest Orlando Lawrence Berkeley National Laboratory [9] has an inlet temperature between 15 and 45°C, whereas the case study presented in [10] has an inlet temperature of 50°C. Other advantages are the considerable fan energy saving and a lower noise level. The main drawback of liquid-cooled systems is the introduction of liquid within the data center and the potential damage that a failure can cause. In 2022, Bitcoin mining company Mint Green to deliver an innovative low-carbon mining waste solution to heat the City of North Vancouver, BC [11]. Production of both bitcoin and usable thermal energy positions the Digital Boilers to be the cost-leading low-carbon heating technology. In MintGreen design, the hash boards have been placed radially in a bell month-like chamber, submerged fully in the single-phase dielectric fluid. The cold coolant will flow in the chamber along the central line, be heated by the radial located hash boards, and then be collected circumferentially. Wisemining's digital boiler designed for residential mining and heating [12] includes a tank container design for ASIC and a 200-liter water tank with two heat exchangers. This product was designed based on two-phase immersion cooling concepts. The drawbacks of two-phase liquid are the coolant is expensive and highly leakable; The condensation of two-phase fluid in the inner tube driven by thermosiphon has a comparatively lower efficiency than that driven by forced convective heat transfer; The position of the miner must be lower than the coil in the water tank, which needs more footprint of this product.

Hybrid liquid-cooled systems are defined by the integration of the direct-to-chip liquid cooling of some high heat density components such as CPUs and DIMMs by microchannel flow [8, 13] or cold-plate heat exchangers [9, 14], with the air cooling of the rest of low heat density components. A 15 kW IBM rack as a typical example of a hybrid liquid cooled system has been tested [3, 15] under different ambient temperatures. This chiller-less rack can work under inlet coolant temperature of up to 45°C. In the summer season, its IT load of 13.16 kW can be maintained under 0.442 kW cooling power consumption. The average cooling power for a full year could be expected to be below 3.5%.

2. Objectives

Doubtless to say, the implementation of waste heat recovery measures in the crypto mining industry to building or district heating, absorption cooling, Rankine Cycle, Organic Rankine Cycle (ORC), biomass co-location, and desalination for clean water production, can tremendously improve financial performance and reduce CO_2 emissions. Nevertheless, the main impediment to introducing this technology in the data center is the low quality of the heat produced from electronic devices, despite the large quantity. The quality of heat is bound by the upper-temperature limits of electronics, which in most cases remain below 85°C. Air cooling is limited by the thermal properties of air which require co-located siting, making the retrofit into the existing data center impossible. The temperature difference across the inlet and outlet of the hybrid air portion is low due to the water loop removing most of the heat, thus the air cooler portion can operate at a low capacity and the water portion should be targeted for heat recovery. Two-phase immersion cooling is restricted by its characteristic of the saturated temperature-pressure curve which must be realized by a complicated evaporation-recondensation structure. As far as the grade/quality of the heat recovered by the above three approaches is concerned, air cooling typically results in outlet temperatures of 25 to 35°C, while outlet temperatures of up to 60°C are possible with single-phase immersion and hybrid liquid cooling [16]. The max. Temperature of two-phase immersion cooling relies on the selection of two-phase coolant and max. Case temperature of chips. 3 M Novec 7100 with 61°C boiling point @ 1 atm is the most common working fluid in two-phase open bath immersion cooling. Even though Marcinichen [17] reports that high temperatures as 60–70°C for liquid-cooled systems and 70–80°C for two-phase cooling systems provide higher waste heat quality and open up a wide range of waste heat reuse opportunities. These temperatures were achieved in the complex direct on-chip evaporator by HFC134a or HFO1234ze under a working pressure of approx. 15 bar. Kuncoro [18] reports that the temperature differential between chips and coolant can be reduced up to 91.3% by the replacement of air cooling with single-phase oil cooling. It not only reveals the reason liquid cooling is a better option to harvest the high-grade thermal energy but also points out the direction to improve it continuously.

In this article, the prototype of a digital boiler has been designed and built that is helping reduce greenhouse gas emissions by repurposing the heat produced by their ASICs. The miner is cooled by spraying coolant, with intend to enhance heat transferring and consequently elevated coolant outlet temperature for high-grade heat recovery. The concept of PUE based on energy has been redefined accordingly regarding the exergy in the reclaimed heat, catering to the demands of reasonable performance evaluation of data center with thermal energy reuse. Experimental testing validated the exergy efficiency of this innovative design.

3. Design of digital boiler

3.1 Spray cooling mechanism-theory

Why spray cooling has been selected as an alternative to extensively used immersion cooling and how to determine the parameters of spray cooling are the challenges in this design. Tracking the motion trajectory of the coolant liquid out of the sprayer, spreading out of the sprayer holes, receding due to viscous effects, splashing by droplet collision, stationary film generating by fluid viscosity on the solid surface, liquid file flowing by liquid momentum and gravity, and liquid flooding and draining will happen in sequence [19, 20]. The interfacial flow of coolant liquid film on the heat sink surface determines the flow pattern of film, through which heat transfer will happen. The flow patterns of droplets out of the sprayer depends basically on their outlet velocity, which are single droplet, droplet train, and droplet burst in the sequence of velocity increments. The impinging momentum of droplets onto the film attached to the solid surface coupled with the film flow pattern are the two main factors of the heat transfer mechanism of spray cooling. In this study, the droplet flow is designed as the droplet train flow- fresh droplets continuously impact the surface at a certain frequency. To investigate the heat transfer of spray cooling from this aspect, Soriano et al. [21] found that the decisive factor to achieve optimal cooling performance is to let the film velocity not be disturbed by the adjacent droplet streams. Zhang et al. [22, 23] further proved that both impact spacing and impingement pattern affect local and global cooling performance on the hot surface. The droplet train impingement among various impingement patterns is the best one for the highest thermal performance, which has also been recommended by Gao [24] after comparing the circular jet impingement cooling with droplet train impingement. As an advanced methodology [25], droplet trains broken by piezoelectric nozzles from more groups of jet flow can make cooling heat flux up to ~170 W/cm^2 with a nozzle diameter of 25 μm.

In the field of Fluid Dynamics, Nusselt number (N_u) is the ratio of convective to conductive heat transfer at a boundary in a fluid, by which the convective heat transfer coefficient could be calculated.

$$N_u = hL/k \tag{1}$$

where h is the convective heat transfer coefficient of the flow, L is the characteristic length, k is the thermal conductivity of the fluid.

In the single-phase regime, Rybicki and Mudawar [26] proposed the correlation for dielectric PF-5050 spray, which is

$$N_u = 4.7{R_e}^{0.61}{P_r}^{0.32} \tag{2}$$

in which R_e is Reynolds number and P_r is Prandtl number.

As the heat transferring mechanism of immersion cooling, N_u of an isothermal flat plate at a specified temperature in the free stream flow can be calculated by [27, 28]:

$$\text{For Laminar flow } (R_e < 5\text{x}105) : N_u = 0.664{R_e}^{0.5}{P_r}^{0.33} \tag{3}$$

Refer to the quantitative comparison in **Figure 1** with constant P_r as 95 and R_e varying from 0 to 4500, N_u of spray flow from Eq. (2) is far higher than that of laminar flow in Eq. (3). Refer to Eq. (1), no doubt to say that h in spray flow will be much higher than that in immersion cooling. In this case, to get the optimal convection heat transfer coefficient coupled with minimal pressure drop and coolant distribution, a series of testing has been conducted to get the best diameter and distribution of spraying holes. 0.6 mm spraying hole has been proved to be a good option to achieve the droplet train flow and can wet the surface of the heat sink effectively.

Figure 1 presents the comparison of N_u between single-phase spray and single-phase immersion under R_e number varying from 0 to 4500. N_u will be 200 for immersion cooling, and 3100 for spray cooling with R_e number 4000. It means under the same access liquid velocity, spray cooling can create much higher heat transferring coefficient than immersion cooling.

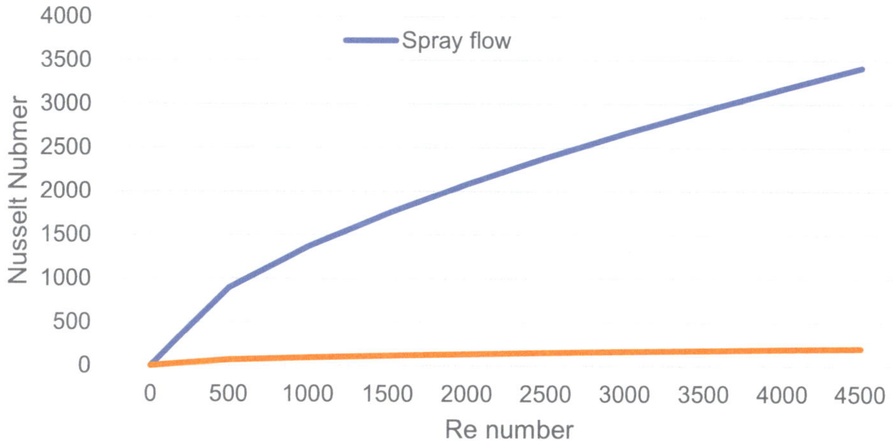

Figure 1.
Comparison of N_u between single-phase spray and single-phase immersion under R_e number varying from 0 to 4500.

3.2 Spray cooling mechanism-design

Figure 2 presents a schematic diagram of dielectric coolant being sprayed on the hash boards and heat sinks. The noisy fans in the air-cooled design have been removed and replaced by a liquid sprayer in the liquid cooling. The miner with the original heat sinks designed for air-cool was installed vertically in the reservoir, then the coolant was sprayed out of the sprayer from the top and access to hash boards and heat sinks. The coolant flush on the surface of the heat sink and drained automatically by gravity. The liquid film on the surface of the heat sink can be kept at its minimal thickness due to the excellent drainage and forceful flushing momentum. The surface area of the front and back heat sink indicates the different heat loads on the front and back side of the hash board, the spraying flow to them is distributed accordingly. The spraying holes on the top sprayer have been pinpointed to the heat sinks and their spatial distribution has been arranged based on the heat load distribution which can guarantee enough wetting on the surface of the heat sink for the best thermal performance with the least coolant consumption.

3.3 Digital boiler-design

In **Figure 3**, a modular heat recovery system by the name of Digital Boiler has been developed for hot water heating. This modular can be extended or combined to be an array providing higher heat capacity. This modular digital boiler includes four major kits, i.e. miner liquid cooling chassis, water tank kit, water replenishing kit, and dry cooler. The thermal energy generated from the miner including three hash boards (WhatsMiner M30S) will heat the dielectric coolant (7.3 L coolant charged) in the enclosure and then be suctioned into the pump (a DC permanent magnet pump). The heated dielectric coolant out of the pump will flow into the line connected with the

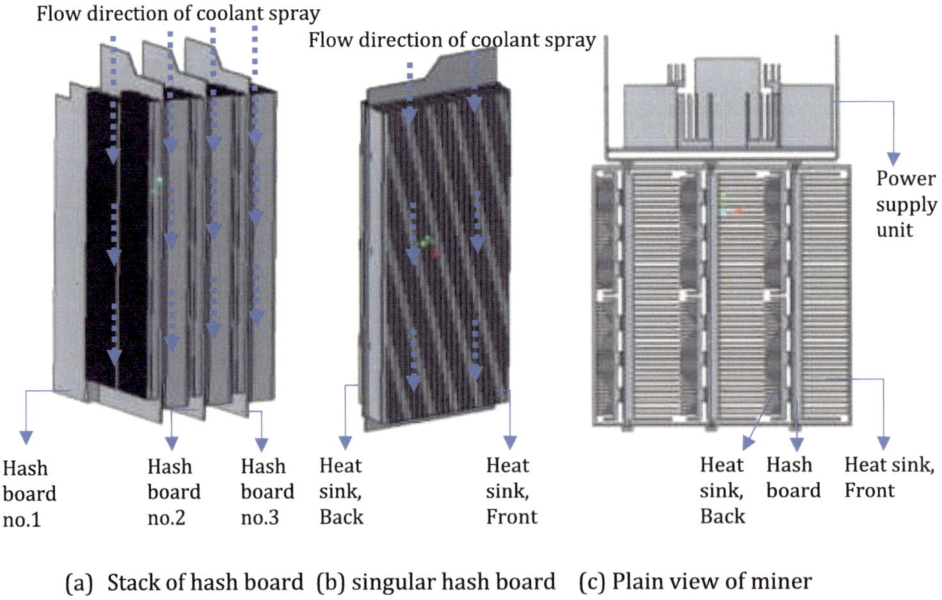

Figure 2.
Schematic diagram of coolant spraying design. (a) Stack of hash board, (b) singular hash board, and (c) plain view of miner.

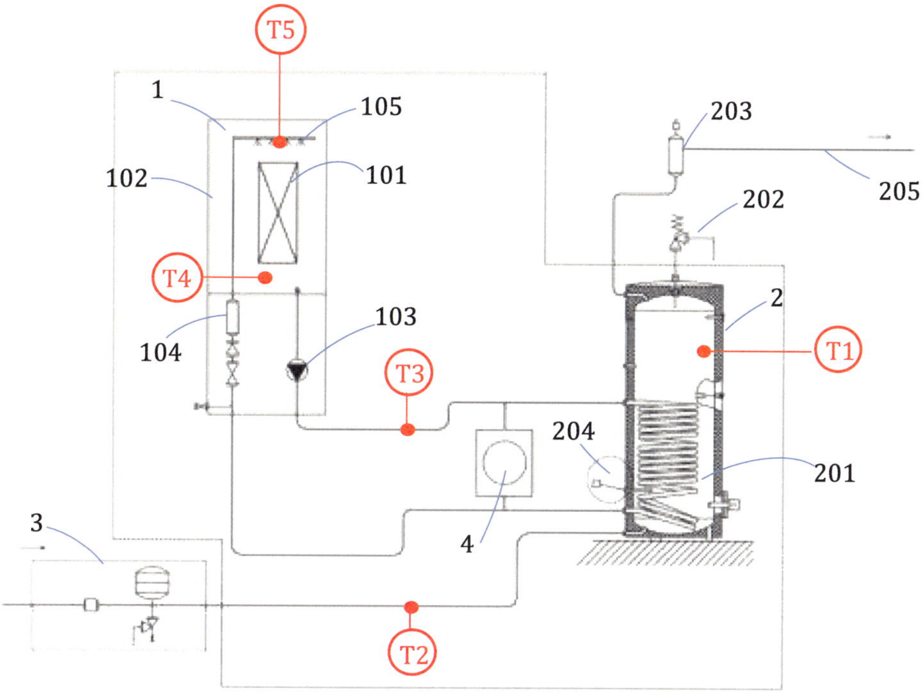

Figure 3.
Typical configurations of a digital boiler. 1 mining liquid cooling chassis, 101 miner, 102enclosure, 103 pump, 104 Filter,105 sprayer, 2 water tank kit,201 Sprial coil,202 relief valve,203 air purger, 204 liquid level controller, 205 supply water line,3 water replenishing kit, 4 dry cooler.

spiral coil in the water tank (a 190 L insulated tank). The internal volume of the water tank would be filled with fresh water from the water replenishing kit. Hot water in the water tank will be heated by coil and pushed out to the supply water line. In the supply water line, the air purger will discharge the air released from water into the ambient. The thermostat can maintain the water temperature in the water tank by controlling the bypass flow through the dry cooler partially or fully in response to the load requests. Thermocouples connected with the datalogger have been put at the positions shown in **Figure 3** for testing purposes. **Figure 4** indicates the main components and their installation of a prototype. **Figure 5** is the image of a real digital boiler, which has the exact compatible geometrical size and connections with the current electric/gas water heater. **Table 1** shows the main parameters of Digital Boiler.

4. Performance of digital boiler

As a key parameter of heat recovery, the max. Water temperature has been studied under the steady, reliable, and continuous operation of miners. **Figure 6** presents the variation of power, average temperature of hash boards and frequency of hash board with different spraying coolant temperature. It can be found in **Figure 6** that under a certain spraying flowrate, the miner power consumption, hash rate, and frequency have a very limited rising with the increase of dielectric coolant temperature (T4 in **Figure 3**). It implies that the miner can deliver a constant hashing rate in parallel with

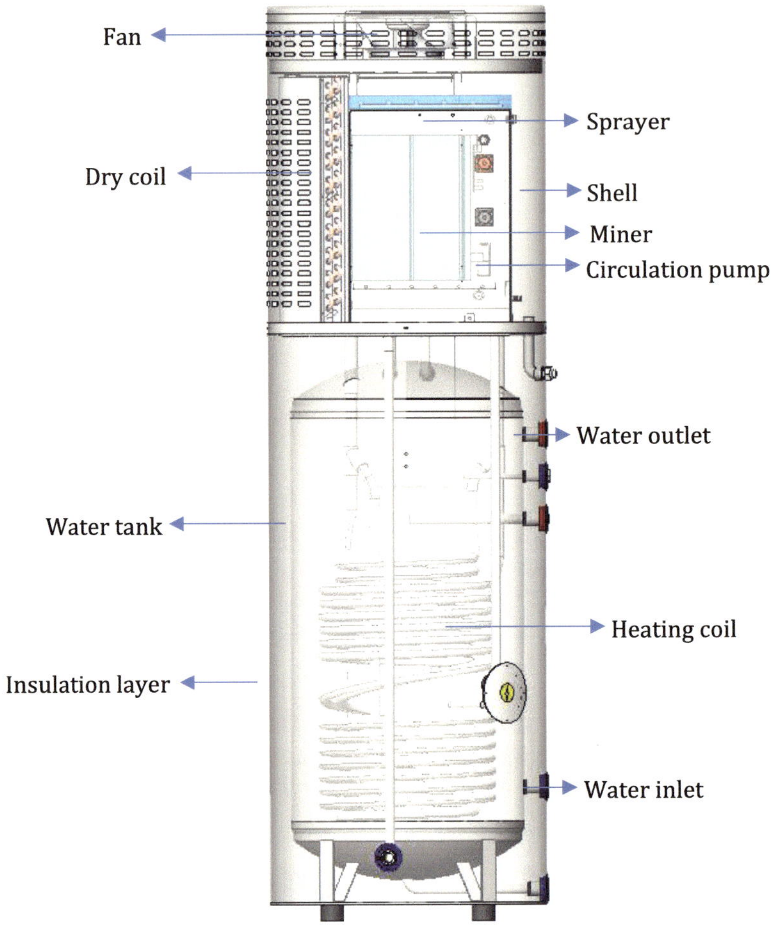

Figure 4.
Drawing of digital boiler.

stable thermal output. The average temperature of hash boards read from Whatsminer-M3x&M5x firmware has a linear trend with the dielectric coolant temperature. At the 45°C spraying dielectric coolant temperature (T5 in **Figure 3**), the temperature differential between the hash board and dielectric coolant is 10.17°C; At the max. Spraying dielectric coolant temperature of 65°C, the temperature differential between the hash board and the coolant is 8.50°C. Lower temperature differential at higher dielectric coolant temperature can be attributed to the viscosity reduction under higher spraying dielectric coolant temperature, which is a positive factor to elevate the dielectric coolant temperature further. To avoid triggering the auto frequency throttling mechanism for internal temperature protection [29], spraying coolant temperature is set as 65°C with safety tolerance for this miner without sacrificing its reliability.

Figure 7 reveals the trend of spray coolant temperature (T5 in **Figure 3**) and hash board temperature as the function of coolant flowrate. The temperature differential between spraying coolant and hash board will reduce with the increase of coolant

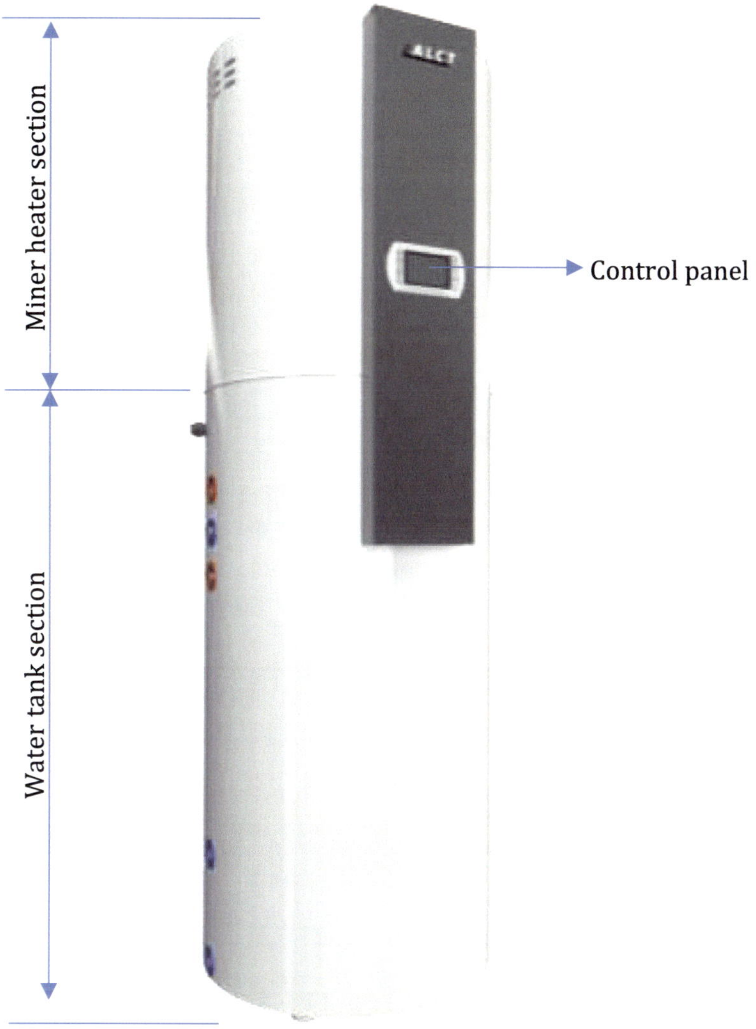

Figure 5.
Real image, and sections of digital boiler.

Height (mm)	Diameter (mm)	Weight (kg)	Nominal volume (L)	Water inlet/ out (inch)	Insulation layer (mm)	Electrical energy input (kW)
1950	600	114	190	¾ NPT	50.8	3.4

Table 1.
Digital boil performance as a residential water heater.

flow. For example, the temperature differential between spraying coolant and the hash board no.2 is 10°C at the flow rate of 25 L/min and decreased to 6°C at the flow rate of 45 L/min. It can be attributed to the compound effects of viscosity reduction under higher temperatures and the flush velocity increase with more coolant. This trend implies a possible approach to elevate the coolant temperature further. In addition, it can be found that with the increased coolant flow to 45 L/min, max. 70°C

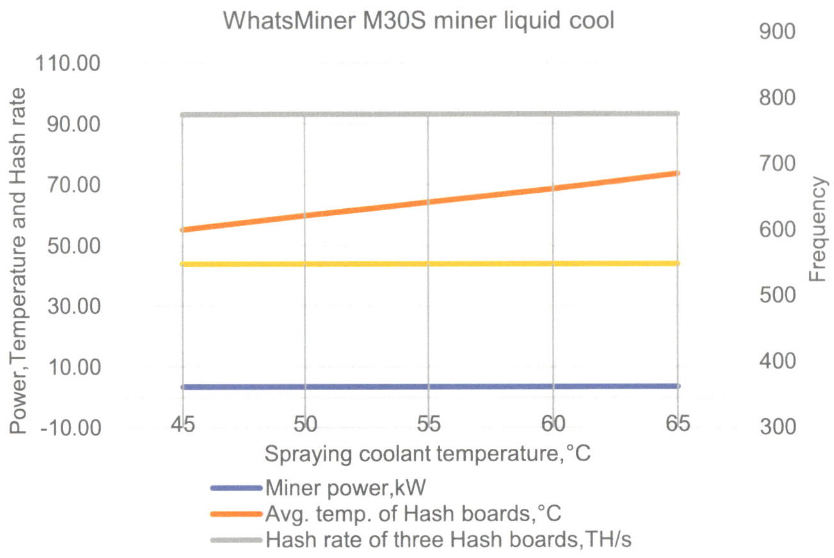

Figure 6.
Variation of power, average temperature of hash boards and frequency of hash board with different spraying coolant temperature.

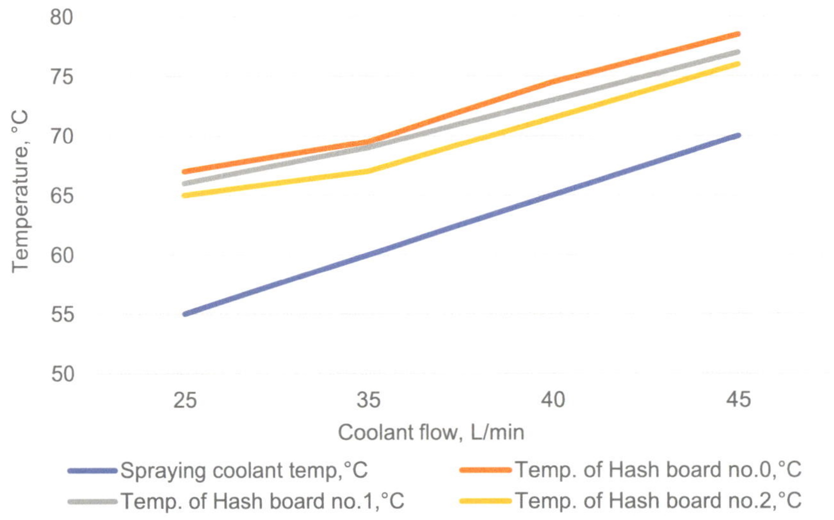

Figure 7.
Variation of hash boards' temperature with different spraying coolant flow rate.

coolant temperature can be achieved with a hash board temperature under 80°C due to the minimized temperature differential by enhanced spraying momentum. This high coolant temperature can not only provide high exergy output for high-grade heat recovery but also provide the capability to kill the legionella in the water system within minutes. To minimize the legionellosis risk for building water systems has been defined compulsorily by ASHARE and CDC as a national code [30–33]. Reviewing the maximum temperature achieved in previously published single-phase dielectric liquid cooling solutions [7–9, 34, 35], 50°C can be considered a record that cannot meet the primary safety requirements of a building water system.

Reviewing the evolution of district energy technologies [36], the operation temperature categorized by International Renewable Energy Agency is dropping from approx. 200°C of First-generation district heating which was based on the steam system, and transferred by steam pipes in concreted ducts during Y1880–Y1930, to the range of 50–70°C of Fourth generation district heating which is based on the smart energy during Y2020–Y2050, including an optimum interaction system of sustainable energy sources, intelligent distribution system, two-way energy reservoir, and end consumption. From Marco and quantitative perspective, the renewable share in global district heat will be increased from 8% (30.64EJ) in Y2017 to 77% (270.27EJ) in Y2050, renewable share in electricity will be increased from 25% (95.75EJ) in Y2017 to 86%(301.86EJ) in Y2050 respectively. Refer to global trends in internet traffic, data center workloads, and data center energy use [37, 38], the global energy consumption by the data center in Y2030 will be 11.52EJ, which could provide 9.4% district heat load to Forth generation district heating directly as renewable energy, if the medium temperature reclaimed from datacenter into district heating network could catch up the range of 50–70°C. Regarding the transmission network losses, output medium temperature from the data center higher than 60°C can be considered as the bottom line to ensure the seamless integration with the Fourth general district heating network. The infrastructure of district heating can be merged with that of the data center as an integrated energy complex. From the micro perspective, the heat pump is one of the most energy-efficient and environment-friendly options to boost the low-grade thermal energy from the data center, COP of the heat pump booster under temperature lift 45°C can exceed 5.6 with some low GWP synthetic refrigerants, i.e. R1234ze and R1234zd. Their optimal working range to achieve high COP is 55–65°C evaporation temperature [39]. ORC is a thermal-electrical recovery for low-grade waste heat. The higher the water temperature is, the higher the cycle efficiency can be achieved. From either macro or micro perspectives, elevating the outlet temperature from the data center is a key to not only determining the system efficiency technically but also impacting the capital investment and revenue return economically. 60°C as the medium outlet temperature of data center heat recovery can be considered the reference temperature technically and financially.

5. Exergy efficiency analysis

The exergy [40] refers to the availability or quality of a thermodynamic system to a specified reference and is related to the first and second laws of thermodynamics. The availability of a thermal system is zero when in balance with the reference conditions. The physical exergy definition is given in Eq. (4) where h_i and h_o refer to the specific enthalpies and s_i and s_0 refer to the specific entropies of ith point and dead state conditions and T_0 is the dead state or ambient temperature. The exergy factor of mechanical energy and electrical energy is 1.0 [41].

$$ex_i = h_i - h_0 - T_0 \times (s_i - s_0) \tag{4}$$

The curve in **Figure 8** is the specific exergy of hot water under the condition of 1 bar pressure and variable temperature. The specific exergy at 50°C hot water (Point 1) is 4.19 kJ/kg, which is half of the specific exergy at 61°C (Point 2). It means the exergy recovered in this design is twice as much as the exergy in the current commercial liquid immersion system. The slope of this curve indicates the increasing ratio

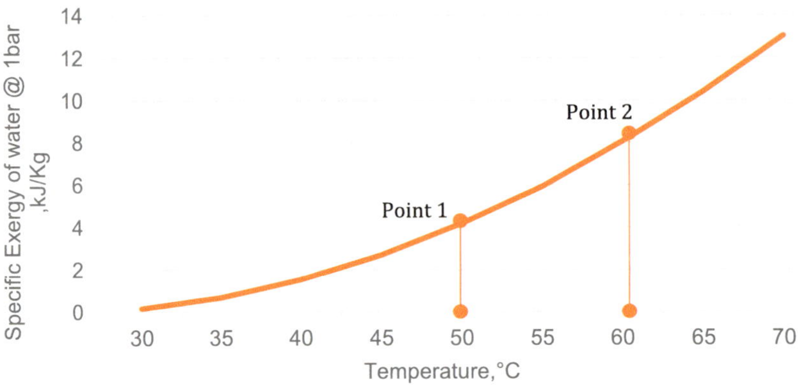

Figure 8.
Exergy in hot water under different temperatures @ 1 atm.

of exergy at the higher temperature is much larger than that at the lower temperature. This characteristic reveals the essence to pursue high-grade heat recovery.

Power usage effectiveness (PUE), a concept based on the first law of thermodynamics, is a ratio that describes how efficiently a computer data center uses energy; Specifically, how much energy is used by the computing equipment in contrast to cooling and other overhead that supports the equipment, which was published in 2016 as a global standard under ISO/IEC 30134-2:2016. An ideal PUE is 1.0, given that Non-IT Facility Energy is zero, refer to Eq. (5). Anything that is not considered a computing device in a data center (e.g., lighting, cooling, etc.) falls into the category of facility energy consumption.

$$PUE_{en} = \frac{TotalFacilityEnergy}{ITEquipmentEnergy} = 1 + \frac{NonITFacilityEnergy}{ITEquipmentEnergy} \tag{5}$$

Since Total facility energy and IT equipment Energy are supplied in the form of electrical energy, the exergy factor of electrical energy is 1. Eq. (5) can be recast as follows:

$$PUE_{ex} = \frac{TotalFacilityExergy}{ITEquipmentExergy} = 1 + \frac{NonITFacilityExergy}{ITEquipmentExergy} \tag{6}$$

Regarding the exergy in the reclaimed thermal energy, the Non-IT Facility Exergy can be offset partially. The definition of PUE_{ex} in Eq. (6) involving the heat recovery will be

$$\begin{aligned} PUE_{ex} &= \frac{TotalFacilityExergy - Exergyinreclaimedheat}{ITEquipmentExergy} \\ &= 1 + \frac{NonITFacilityExergy - Exergyinreclaimedheat}{ITEquipmentExergy} \end{aligned} \tag{7}$$

The exergy efficiency of the heat reclaim system will be defined as

$$E_{re} = \frac{Exergyinreclaimedheat}{Totalfacilityexergy} \tag{8}$$

Total facility energy (kW)	IT equipment energy (kW)	PUE_{en}	Total facility exergy within 1 hour (kJ)	Exergy in reclaimed heat within 1 hour (kJ)	IT Equipment Exergy within 1 hour (kJ)	PUE_{ex}	E_{re} (%)
3.40	3.30	1.03	12225.60	997.13	11880.00	0.95	8.16

Table 2.
Exergy calculation of heat reclaim.

Table 2 lists the testing results and derivative calculation of this prototype. The miner consumes 3.3 kW electrical energy at its nominal hash rate. The energy consumption of the circulation pump is 100 W. PUE refer to Eq. (5) would be 1.03. Total input exergy from electrical energy within 1 hour is 12225.6 kJ. Regarding the heat recovered from the miner, the specific exergy in the hot water at 63.23°C@1 atm is 9.62 kJ/kg obtained in **Figure 8**. The total reclaimed exergy at the hot water supply flow of 103.5 kg/h during its first hour of running. Based on Eqs. (6) and (8), PUE_{ex} and E_{re} will be 0.95 and 8.16% respectively. The conventional PUE_{en} is 1.03, in contrastPUE_{ex} is less than one with consideration of reclaimed exergy. As far as the rate of energy reuse is concerned, PUE_{ex} can be considered as a better index than PUE_{en} due to its characteristic energy-quality measures.

6. Conclusions

A water heater with the heating element as a bitcoin miner has been designed and built, in which the single-phase dielectric coolant has been sprayed on the hash boards and coupled heat sinks of the miner in steady conventional submerge. By avail of the spray momentum, the resultant heat transferring has been enhanced and validated by increased N_u in theory and up to 70°C outlet coolant temperature in the test. The temperature differential between coolant and hash board is controlled as lower as 10°C, by which the energy grade to be recovered from either centralized or decentralized mining operations has been elevated largely. From the perspective of the first law of thermodynamics, the liquid spray cooling circulation can extract the heat from miners and transfer the energy to the exterior with minimal losses. From the perspective of the second law of thermodynamics, the quality of thermal energy reclaimed in the miners, as the monotonically increasing function of the output medium temperature, has been elevated to the extent that the district heating system can be integrated seamlessly with datacentre cooling system without extra infrastructure investments, and critical hygienic codes have also been fulfilled completely. PUE, a metric used to determine the energy efficiency of a data center, has been reconsidered and redefined in the exergy flow analysis, rather than the energy flow previously. From the testing results of the prototype, PUE_{en} is 1.03 and PUE_{ex} is 0.95. The discrepancy between them tells the influence of reclaimed useful energy, which must be considered with the prevalence of energy-saving thinking in the data center industry.

This study is the initial part of the synthesis energy system integrating digital energy (energy reclaimed from extensive digital industry), fossil energy, and sustainable energy. Regarding the stable and unidirectional output characteristics of 7x24h operation in the datacentre, the reclaimed thermal energy can work as the basic component of the district heating load, which can ease the tensions in the network caused by the fluctuating sustainable energy like wind, and solar energy. With the quickly developing electronic industry, the upper-temperature limits of chips tend to rise continuously. The electrical power generator based on ORC would be a good solution to realize the electrical-thermal-electrical close loop in the data center. The advantage of this thermal-electrical transition is the saving of access pipework to the current heating system. However, this thermal-electrical recovery is more applicable to large-scale scenarios. Considering the high start temperature of ORC, more studies are required to suit the 60–70°C heat sources.

Abbreviations

ANSI	American National Standards Institute
ASHRAE	American Society of Heating, Refrigerating and Air-Conditioning Engineers
CDC	Centers for Disease Control and Prevention
COP	Coefficient of Performance
DC	Direct Current
GWP	Global Warming Potential
IEC	International Electrotechnical Commission
ISO	International Organization for Standardization
IT	Intelligent Technology
Nu	Nusselt number
ORC	Organic Rankine Cycle
Pr	Prandtl number
PUE	Power Usage Effectiveness
Re	Reynolds number

Author details

Nan Chen[1,2]*, Yunshui Chen[2] and He Zhao[3]

1 Advanced Liquid Cooling Technologies, Greer, USA

2 Airsys Cooling Technologies, Greer, USA

3 Airsys Singapore Pte. Ltd, Singapore

*Address all correspondence to: brad.chen@advancedliquidcooling.com

References

[1] Cristina C. Bitcoin consumes 'more electricity than Argentina' [Internet]. 2021. Available from: https://www.bbc.com/news/technology-56012952. [Accessed: June 24, 2022]

[2] Gibbons L, Persoons T, Alimohammadi S. Techno-economic and sustainability analysis of potential cooling methods in Irish data Centres. Journal of Electronics Cooling and Thermal Control. 2021;**10**(3):35-54. DOI: 10.4236/jectc.2021.103003

[3] Ljungqvist HM, Mattsson L, Risberg M, Vesterlund M. Data center heated greenhouses, a matter for enhanced food self-sufficiency in sub-arctic regions. Energy. 2021;**215**:119169. DOI: 10.1016/j.energy.2020.119169

[4] Enachescu MS Closed Loop Cryptocurrency Mining in Alberta [Internet]. 2019. Available from: https://prism.ucalgary.ca/bitstream/handle/1880/111108/2019%20Closed%20Loop%20Cryptocurrency%20Mining%20in%20Alberta.pdf?sequence=1&isAllowed=y. [Accessed: June 24, 2022]

[5] Agrodomes company website. Agrodomes: Home [Internet]. 2022. Available from: https://agrodomes.com/. [Accessed: June 24, 2022]

[6] United American Corp. United Blockchain Corp Announces International Request for Proposal (RFP) For Its Proposed BlockchainDomes in Quebec for Large-scale mining operations [Internet]. 2018. Available from: https://www.globenewswire.com/news-release/2018/01/24/1304376/0/en/United-Blockchain-Corp-Announces-International-Request-for-Proposal-RFP-For-Its-Proposed-BlockchainDomes-in-Quebec-for-Large-Scale-Mining-Operations.html. [Accessed: June 24, 2022]

[7] Kim MH, Ham SW, Park JS, Jeong JW. Impact of integrated hot water cooling and desiccant-assisted evaporative cooling systems on energy savings in a data center. Energy. 2014;**78**:384-396. DOI: 10.1016/j.energy.2014.10.023

[8] Zimmermann S, Meijer I, Tiwari MK, Paredes S, Michel B, Poulikakos D. Aquasar: A hot water-cooled data center with direct energy reuse. Energy. 2012; **43**(1):237-245. DOI: 10.1016/j.energy.2012.04.037

[9] Coles HC, Steve EG. Direct liquid cooling for electronic equipment [Internet]. 2014. Available from: https://eta-publications.lbl.gov/sites/default/files/direct_liquid_cooling.pdf. [Accessed: June 24, 2022]

[10] Chi YQ, Summers J, Hopton P, Deakin K, Real A, Kapur N, et al. Case study of a data center using enclosed, immersed, direct liquid-cooled servers. In: 2014 Semiconductor Thermal Measurement and Management Symposium (SEMI-THERM). San Jose, CA USA: SEMI-THERM; 2014. pp. 164-173. DOI: 10.1109/SEMI-THERM.2014.6892234

[11] Jesse W. North Vancouver to Be World's First City Heated by Bitcoin [Internet]. 2021. Available from: https://www.nasdaq.com/articles/north-vancouver-to-be-worlds-first-city-heated-by-bitcoin-2021-10-14. [Accessed: June 24, 2022]

[12] Wisemining company website. Meet Sato [Internet]. 2022. Available from: https://www.wisemining.io/product. [Accessed: June 24, 2022]

[13] Capozzoli A, Primiceri G. Cooling systems in data centers: State of art and

emerging technologies. Energy Procedia. 2015;**83**:484-493. DOI: 10.1016/j.egypro.2015.12.168

[14] Asetek, Inc. Global leader in liquid cooling Solutions [Internet]. 2022. Available from: https://www.asetek.com/liquid-cooling/. [Accessed: June 24, 2022]

[15] Gaynes M, Simons R, Schmidt R, Chainer T. Server liquid cooling with chiller-less data center design to enable significant energy savings. In: 2012 28th Annual IEEE Semiconductor Thermal Measurement and Management Symposium (SEMI-THERM). San Jose, CA USA: SEMI-THERM; 2012. pp. 212-223. DOI: 10.1109/STHERM.2012.6188851

[16] Koronen C, Åhman M, Nilsson LJ. Data centers in future European energy systems—Energy efficiency, integration, and policy. Energy Efficiency. 2020; **13**(1):129-144. DOI: 10.1007/s12053-019-09833-8

[17] Marcinichen JB, Olivier JA, Thome JR. On-chip two-phase cooling of data centers: Cooling system and energy recovery evaluation. Applied Thermal Engineering. 2012;**41**:36-51. DOI: 10.1016/j.applthermaleng.2011.12.008

[18] Kuncoro IW, Pambudi NA, Biddinika MK, Widiastuti I, Hijriawan M, Wibowo KM. Immersion cooling as the next technology for data center cooling: A review. Journal of Physics: Conference Series. 2019; **1402**(4):044057. DOI: 10.1088/1742-6596/1402/4/044057

[19] Finch A. Mechanics of Spray Cool Direct Spray [Internet]. 2009. Available from: https://www.parker.com/literature/Gas%20Turbine%20Fuel%20Systems%20Division/TMS%20Microsite%20Literature%20files/PH_WP_Mechanics_of_SprayCool_direct_spray.pdf. [Accessed: June 24, 2022]

[20] Gao X, Li R. Spray impingement cooling: The state of the art. In: Advanced Cooling Technologies and Applications. Vol. 5. London: IntechOpen; 2018. pp. 27-51. DOI: 10.5772/intechopen.80256

[21] Soriano GE, Zhang TL, Alvarado JL. Study of the effects of single and multiple periodic droplet impingements on liquid film heat transfer. International Journal of Heat and Mass Transfer. 2014; **77**:449-463. DOI: 10.1016/j.ijheatmasstransfer.2014.04.075

[22] Zhang TL, Alvarado JL, Muthusamy JP, Kanjirakat A, Sadr R. Numerical and experimental investigations of crown propagation dynamics induced by droplet train impingement. International Journal of Heat and Fluid Flow. 2016;**57**:24-33. DOI: 10.1016/j.ijheatfluidflow.2015.10.003

[23] Zhang TL, Alvarado JL, Muthusamy JP, Kanjirakat A, Sadr R. Heat transfer char,acteristics of double, triple and hexagonally-arranged droplet train impingement arrays. International Journal of Heat and Mass Transfer. 2017; **110**:562-575. DOI: 10.1016/j.ijheatmasstransfer.2017.03.009

[24] Gao X, Li R. Impact of a drop burst flow on a film flow cooling a hot surface. International Journal of Heat and Mass Transfer. 2018;**126**:1193-1205. DOI: 10.1016/j.ijheatmasstransfer.2018.06.042

[25] Chen H, Cheng WL, Peng YH, Zhang WW, Jiang LJ. Experimental study on optimal spray parameters of piezoelectric atomizer-based spray cooling. International Journal of Heat

and Mass Transfer. 2016;**103**:57-65. DOI: 10.1016/j.ijheatmasstransfer.2016.07.037

[26] Rybicki JR, Mudawar I. Single-phase and two-phase cooling characteristics of upwardfacing and downward-facing sprays. International Journal of Heat and Mass Transfer. 2006;**49**:5-16. DOI: 10.1016/j.ijheatmasstransfer.2005.07.040

[27] Holman JP. Heat Transfer. 7th ed. Southern Methodist University, McGraw Hill Book Company; 1989. p. 736

[28] Delgado San Román F, Fernández Diego I, Urquiza Cuadros D, Mumyakmaz B, Unsal A. Fluid-thermal analysis of the cooling capacity of a commercial natural ester in a power transformer. Renewable Energy and Power Quality Journal (RE&PQJ). 2013; **1**(11):1158-1163. DOI: 10.24084/repqj11.564

[29] Shenzhen MicroBT Electronics Technology Co., Ltd. Whatsminer M30S Manual [Internet]. 2019. Available from: https://www.zeusbtc.com/manuals/user-manuals/WhatsMiner-M30S-Manual.pdf. [Accessed: June 24, 2022]

[30] ASHRAE. ANSI/ASHRAE Standard 188-2018. Legionellosis: Risk management for building water systems establishes minimum legionellosis risk management requirements for building water systems. 2018

[31] ASHRAE Guideline 12-2000. Minimizing the Risk of Legionellosis Associated with Building Water Systems

[32] U.S. Department of Health and Human Services Centers for Disease Control and Prevention (CDC). Guidelines for environmental infection control in health-care facilities [Internet]. 2003. Available from: Guidelines for environmental infection control in health-care facilities (cdc.gov). [Accessed: June 24, 2022]

[33] U.S. Department of Energy. Water heater test procedure rulemaking: development testing preliminary report [Internet]. 2013. Available from: https://www1.eere.energy.gov/buildings/appliance_standards/pdfs/water_heater_test_report_111413.pdf. [Accessed: June 24, 2022]

[34] GRC. Powerful immersion cooling technology in an incredibly compact, plug-n-play package [Internet]. 2021. Available from: https://www.grcooling.com/wp-content/uploads/2021/03/grc_data_sheet_iceraq_micro.pdf. [Accessed: July 5, 2022]

[35] Submer. Micropod technical specs [Internet]. 2022. Available from: https://submer.com/micropod/. [Accessed: July 5, 2022]

[36] Federal Ministry for the Environment, Nature Conservation and Nuclear Safety. Integrating low-temperature renewables in district energy systems: Guidelines for policy makers [Internet]. 2021. Available from: https://irena.org/-/media/Files/IRENA/Agency/Publication/2021/March/IRENA_District_Energy_Systems_2021.pdf. [Accessed: July 30, 2022]

[37] Benjamin KS, Chukwuka GM, Paul U. Making the internet globally sustainable: Technical and policy options for improved energy management, governance and community acceptance of Nordic datacenters. Renewable and Sustainable Energy Reviews. 2022;**154**: 1-20. DOI: 10.1016/j.rser.2021.111793

[38] Liu Y, Wei X, Xiao J, Liu Z, Xu Y, Tian Y. Energy consumption and emission mitigation prediction based on data center traffic and PUE for global data centers. Global Energy

Interconnection. 2020;**3**:272-282. DOI: 10.1016/j.gloei.2020.07.008

[39] Wang RZ, Xu ZY, Hu B, Du S, Jiang L, Wang LW. Heat pumps for efficient low grade heat uses: From concept to application. Thermal Science and Engineering. 2019;**27**:1-15. DOI: 10.11368/tse.27.1

[40] Trusler JPM. T-S diagram for water [Internet]. 2003. Available from: https://www.imperial.ac.uk/media/imperial-college/research-centres-and-groups/thermophysics/Chart-T-s-Water.pdf. [Accessed: June 24, 2022]

[41] Gong M, Wall G. Exergy analysis of the supply of energy and material resources in the Swedish society. Energies. 2016;**9**(9):707. DOI: 10.3390/en9090707

Chapter 4

Steam Heating Conundrum of High-Rise Buildings

Igor Zhadanovsky

Abstract

Steam/vacuum heating systems are employed in thousands of old tall buildings and even relatively new ones. Although considered obsolete, these systems excels any other in simplicity and resilience and can match modern efficiency standards after thorough retrofit. Efficient and simple, the NextGen vacuum heating is suggested for existing steam heating systems retrofit and for new installations that use modern technologies and materials.

Keywords: high-rise, vacuum heating, steam, retrofit, steam heating

1. Introduction

Till 1930 all skyscrapers were built in North America/US (191/188) [1]. Typical office towers and high-rise residential and hotel buildings were not higher than 150 meters [2]. They were heated by steam systems, – the most convenient way to deliver heat to upper floors. Reliable hot water circulators were introduced in 1929, and quickly gained popularity because the cost of heating system installation was reduced drastically. It's a reason why in this article most of the data on steam heated high rise buildings are of US/NYC origin, but proposed solution can be applicable for future high-rise construction worldwide.

In Europe a record number of skyscrapers were completed in the first decade of the 21st century and it appears that this record might be surpassed in the next. Although Europe constructed first supertall skyscrapers, European buildings generally do not exceed 250 meters. Today, the majority of European skyscrapers are office buildings; although in the 21st century there has been an increasingly significant tendency to equalize the quantity of residential and office buildings [2]. The skyscrapers are changing the skylines of high populated cities - **Figure 1** (NYC) and **Figure 2** (Chicago) are good examples.

The elephant swept under the rug – steam heating in high-rise buildings. Swept under the rug is an idiom used for something that has been hidden from the view of others due to embarrassment ... You have all the dirt, but too lazy or lacking in time to find something to pick it up, so you lift the corner of the rug and sweep it under.

Today's consensus on HVAC with the highest efficiency for tall buildings is a "ground source heat pump (HP) combined with hydronic piping, and smaller water source heat pumps for each building zone. One kilogram of water can carry over four

Figure 1.
The downtown New York City skyline.

Figure 2.
Skyscrapers in Chicago.

times more heat than one kilogram of air, while being pumpable and using over 800 times less space. This makes water an ideal heat transfer medium for tall construction" [3].

Apparently, enthusiasm for heat pumps obscures the fact that the best heat transfer media is steam, not water. 1 pound of condensed steam carries more heat than 25 pounds of hot water cooled in radiators by 40°F, no pumping is required. Additionally, in tall buildings every 20th floor of precious space is lost for hot water pumps, which is a huge budget sacrifice. The choice of a HP is also questionable because HP efficiency drops sharply at low outside temperatures, meaning the backup system is required when heating is needed the most. Backup system is either an electrical resistance or Power Plant. HP requires 24/7/365 electricity which is produced from oil or gas at maximum theoretical efficiency of 35%. To circumvent such losses, free "green" electricity from solar and wind is asserted as todays "holy grail" to power HP.

In real life, the iconic "green" program in Germany upsurge electricity cost and was finally exposed as a catastrophic failure [4]. Still the public is mesmerized by futuristic ideas. "The growth of wind and solar power offer the opportunity to reduce the cost of electricity, and certainly reduce the amount of emissions. However, this only accounts for about 3% of the total U.S. energy use or about 9% of electric generation energy. In addition, electric generation from wind is often at the wrong time of the day. While there are many ways of solving these problems, solutions will not be easy or inexpensive. Even if there is a conversion to more electric generation from wind and solar, steam will still be the best means of converting electricity to a useful energy source for heating and process work. It simply means that the boiler will use electricity for its energy source. The flexibility offered by steam and steam generation from boilers will continue to keep the use of steam in our foreseeable future" [5]. Heat and power cogeneration (CHP) efficiency is up to 75% and 24/7 reliability, which makes it an attractive option for new large/tall buildings and campuses.

Pragmatism prevailed over wishful thinking, – according to 2010 reports in NYC "the tradition of steam heat is so strong that even relatively new buildings have been found to be designed and built with steam heating systems" [6]. "28 newly constructed properties, totaling over 7.5 M SF (1SF = 0.0929 M2) of building area, installed some form of steam heat between 2000 and 2010. Most buildings above 50,000 SF still use steam-based space heating. 72.9% of buildings have steam boilers fired by natural gas or fuel oil, while 10% rely on Con Edison's district steam service. In other words, 81.9% of heating systems in large NYC buildings still use steam [7]. The propensity/partiality toward steam heating in large buildings is typical of big cities in the US and in Europe as well [8].

2. Solutions already tested, found inefficient but still employed

The inborn drawback of steam heating is uneven heat distribution. Steam of 2 psi pressure (safety limit) has to push air through multiple air vents every heating cycle in order to get simultaneously into each radiator hundreds feet away. That's why ¨Dead Men¨ meticulously equilibrate boiler capacity, pipes and radiators sizes, pressure drop and heat load for each installation. As of today, efficiency and comfort in the majority of these systems is ruined due to poor maintenance and reckless modifications. It's a reason why steam heated buildings are typically pictured with open windows. Methods to overcome the problem are as follows:

Balancing using air vents [9, 10], orifices [11] and Temperature Regulating Valve (TRV) [12] is time, money and labor intense, resulted a modest 10–14% fuel savings or no savings at all, but still widely employed to supplement building envelope and lighting upgrade.

"Smart' radiator cover" [13] addresses room overheating rather than the whole heat distribution problem, another disadvantage is switching from radiation to convection heating by air which causes a significant decrease in efficiency and comfort.

Variable refrigerant flow (VRF) is today's popular type of HP in which one outdoor unit can be connected to multiple indoor units. Each indoor unit is individually controllable by its user and a variety of unit styles can be mixed and matched to suit individual tenant requirements (e.g. high wall units, cassettes, and ducted units). Recent study concluded that buildings most suitable for MSHP (Multi Split Heat Pump) retrofits are those with high-cost heating fuel such as liquefied petroleum gas (propane), fuel oil and, especially, electric resistance. Buildings with natural gas service are less suitable. Converting from electric resistance to MSHPs saves approximately 30% of annualized energy- related. Converting from oil saved about 4%. Converting from natural gas would cost about 30% more. In the Boston climate, converting from electric resistance to MSHPs is projected to save approximately 50% of annualized energy-related costs. Converting from oil saves about 3%. Converting from natural gas would cost 24% more. VRFs are suitable as primary heat sources in new well insulated construction" [14]. In old buildings (and most steam heated buildings are 50+ years old) a backup heating system is a must to have.

3. Today's best solution

Steam heating conversion into hot water heating (HWH) is considered the best approach for today. The take-away from the 2010 successful steam to HWH retrofit project on a twelve story building at 179 Henry str., NYC were summarized as follows: "Over the years, many of us in the New York City multifamily energy efficiency world have talked about how cool it would be to convert steam-heated buildings to hydronic heating. The problem is not one of will but one of money. Changing the boiler is not the big deal—it's the heating distribution system that is the challenge. ... Plenty of these conversions have been done in the last 20 years in buildings that were gut rehabs. These jobs did not always get the best boilers, or insulation in the walls, but they did get a more efficient heating distribution. The real challenge was to convert a building with steam heat, with tenants in place" [15]. The project required boiler replacement, core drilling the concrete deck floors (12 in all), running and enclosing the new piping and the heating elements. Fuel savings were 33%.

4. Examples of successful steam heating

The efficiency of a thoughtfully retrofitted steam heating system can match today's standards:

- The 16-story 1893 Monadnock Building (Chicago) manages a top Energy Star score of 98 in spite of its age. ... the brick skyscraper has cut electricity and gas consumption by about 33% by weather stripping, improved steam system automation, and the gradual installation of sensor controlled lighting [16]

- The 14-story historic Joseph Vance Building (Seattle) constructed in 1929 and retrofitted in 2006. For economic reasons, the project did not replace the existing steam heating system, but recalibrated it instead. In 2009, the *hot water heating (HWH)* awarded the Vance Building LEED for Existing Buildings (EB) Gold certification. The building also achieved an Energy Star rating of 98 (out of 100) [Kheir [17]].

Besides efficiency, superior comfort can be achieved by steam heating systems even in very old buildings (1850th), – US presidents and their spouses in White House and financial advisors in the US Treasury buildings would not accept anything but the best [18, 19]. Treasury building received a LEED Gold certification in 2011 [20]. These are not tall buildings, but the technology is the same.

To resume:

- There are multiple cases of worn out steam heating system successfully replaced by HWH resulting in a savings ranging up to 40% (usually complementary to a building envelope and windows improvements)

- Many times old steam heating system demonstrated superior comfort and savings in a range of 30–50% thanks to knowledgeable maintenances and tune up

- It's strange, in almost 100 years since the HWH introduction there was no single study of steam heating direct comparison to HWH. There is, though, accidental evidence of similar fuel usage in identical buildings with HWH and well-maintained steam heating.

Public fondness toward HWH is constantly fueled by complaints about noise and low efficiency of the steam heating system which continued functioning even after decades of neglect and scarce maintenance. Maybe it's more rational to find a way to optimize steam heating systems performance rather than proceed with expensive conversions? Many of steam heating buildings are more then 70–80 years old. With average building life of 120 years, such conversion is economically questionable.

5. Evolved solution: nextGen vacuum heating

Vacuum heating, – another kind of steam heating, – was very popular in the 1910th and actually preceded heat pipes concept. Heat Pipe (please, do not mistake it with Heat Pumps) – is the most efficient, resilient, and electricity independent method of heat transfer, employed in NASA spaceships since the 1970th. Basically, it's a closed ends tube under vacuum where working liquid evaporates at one end and releases latent vaporization heat by condensing at the other end.

According to the 100 years old data on steam heating conversions into vacuum heating, reported fuel savings were 30–35% [21]. Instead of pushing air from the system by steam at 2 psi, steam is pulled from the boiler by 10–25"Hg (1"Hg = 0.033421 atm) vacuum at speed of up to 150 mph (1 mph = 0.44704 m/s), which ensures quick and even heat delivery to the farthest radiators in the system. Additionally, thanks to the naturally induced vacuum in the idle cooling down system more heat is sucked up from the boiler after the heating cycle, which would otherwise be lost in the steam systems. At the same time corrosion is reduced because of limited oxygen access into the system. On top of this, steam temperature can be regulated by the vacuum level in the system, so soft comfortable heat can be delivered in warmer weather, rather than being fixed to the 214-218°F range found in steam systems. A good example is the 80+ years old vacuum heating system in the iconic LEED Gold Empire State Building. In the 2009 retrofit, instead of replacing the old vacuum system by hydronic, it was successfully restored to the original design.

The vacuum return systems were widely used to speed up a cold start of steam system heating typically in large and tall buildings. The technology was so conventional that the sale of more than 60,000 vacuum pumps was reported from a single leading supplier since 1921, with most (about 50,000) purchased prior to 1980 [22]. Most of these abundant systems are still around waiting for upgrade/retrofit.

Unlike Heat Pipes, today's vacuum heating is actually a "pseudo" vacuum; the vacuum section of the system is separated from the section under steam pressure by a steam trap behind each radiator(s). These required steam traps present an ongoing maintenance problem. Steam traps on radiators last only 10 years at best, and are often ignored when they fail because of the expense and annoyance of repair. Steam leaking through a single failed trap (out of hundreds in tall buildings, – 6600 in Empire State Building) overloads the vacuum pump, condensate pump, etc. The result is unbalanced, noisy, and very expensive systems. Compared to many other vacuum systems that gave up after a multi-year struggle, Empire State Building definitely excels in steam traps preventive maintenance.

For existing steam/vacuum systems steam traps, this problem can be resolved by converting existing steam/vacuum heating into NextGen vacuum heating system. Observation of vapor and condensate flowing through the transparent plastic piping in the vacuum heating system revealed surprising insights. It turned out, the vacuum system can self-balance quickly and evenly with proper system design [23]. The trick

Effect of	System Comparison	Predicted Consumption (Therms)	Percent Savings
Vacuum distribution	Old Boiler - Single Pipe Steam	1004	
	Old Boiler - Vacuum System	741	26.2%
	New Boiler - Single Pipe Steam	840	
	New Boiler - Balanced Vacuum	500	40.5%
Boiler Upgrade	Old Boiler - Single Pipe Steam	1004	
	New Boiler - Single Pipe Steam	840	16.3%
	Old Boiler - Vacuum System	741	
	New Boiler - Balanced Vacuum	500	32.5%
Total	Old Boiler - Single Pipe Steam	1004	
	New Boiler - Balanced Vacuum	500	50.2%

Table 1.
Result of steam heating retrofit pilot study.

is to prevent steam "short passing" toward the vacuum pump. This has inspired a new "steam traps free", entirely under the vacuum system paradigm [24]. Pilot study confirmed fuel gas savings for two different systems used to heat the same apartment – **Table 1**. Energy efficiency of the original single pipe steam system (~100 years old boiler, piping and radiators) was compared to the same boiler connected to new flat panel radiators by copper/plastic lines; the same comparison was carried later on new regular steam boiler. While boiler upgrade in steam system saved 16.2% energy, 32.5% savings were achieved in vacuum system. Up to 50% in savings was demonstrated by the complete retrofit from steam into a vacuum heating system [25]. In the winter of 2014–2015 results were confirmed.

Steam heating systems in tall buildings are either 2-pipe steam or 2-pipe vacuum heating systems. Without any contempt for the great planning and implementation of the 179 Henry street. NYC project, imagine now that the same building was converted into new vacuum heating system:

- Old boilers, piping and radiators can be salvaged/upgraded/fixed/repaired after a leak tests, steam traps are either left in place or removed

- The only new equipment is a vacuum pump, steam/condensate separator (existing vacuum pump and separator can be employed), sensors and controllers

- The only new piping is the vacuum line on the second-to-last floor ceiling and connecting return lines to the vacuum pump (located either in the basement, on the roof, or designated room on the top floor).

- Plumbing and radiators may be upgraded later on when the building will go into gut rehab.

- Estimated fuel savings - 30-35% according to 100 years old data on steam heating conversions into vacuum heating [21].

This is a minor job compared to the complete replacement of existing steam boilers, piping and radiators by hydronic, on the top of adding mechanical floors (for skyscrapers higher than 20 floors), circulators, PRVs, etc.; not to mention the disturbance of tenants, – the main reason why most retrofits are indefinitely postponed.

For the systems with old piping, the majority of leaks are at supply valves near the radiators and can be fixed. In our experience with three retrofits of residential systems (all are 100+ years old steam heating) no major leak was found in the pipes. Minor leaks from hidden inner-wall piping are inevitable, but the problem can be resolved by converting the steam system into a "vacuum boost" heating system. Here, steam quickly and evenly fills the system under vacuum, raises the pressure to 1–2 psi (to avoid air leakage into the system) and continues to heat the building at low positive pressure till the thermostat is satisfied. Retrofit into "vacuum boost" technology does not require frustrating efforts of leak detection and repair. For "cold" vacuum sustaining test a drop from 20 to 5"Hg in 2 hours is acceptable, compared to a 20–18" Hg drop in 2 hours required by the Vary-Vac – the most popular today's vacuum heating system.

Compared to HWH, leaks in a vacuum heating system are much less of a problem - no water flooding to the lower floors and expensive repairs. The only moving part in the NextGen Vacuum heating system, – vacuum pump –is never exposed to steam and is employed in 5–10 minutes intervals for 1–1.5 hours daily. With a 5–10 thousand hours warranty time and average of 4 month winter season, life expectancy is in a 37–75 years range until the first needed repair/replacement. Condensate is returned into the boiler by gravity; in case of district steam, by a designated pump.

6. Modern technologies prospects for vacuum heating systems in tall buildings

6.1 Plumbing

All existing steam/vacuum heating systems employ heavy steel piping. ESB plumbing contains 50 miles of radiator pipes, including 24 inch (1 inch = 2.54 cm) risers, and 7000 cast iron radiators. Soldered copper tubing was restricted from steam/vacuum heating because rapid heating caused cracks in the soldered joints. Modern ProPress plumbing method employs no soldering, slash installation time and is backed by a 50-year warranty against rust, defects in material and workmanship. Furthermore, the piping diameter/weight is reduced drastically; also reduced is heat loss, the amount of condensate in pipes, and heating time compared to steel piping. In the NGRID study, the 2 ½" steel pipe at the steam boiler exit was replaced by a ¾" copper tube – this resulted in an 11.5 fold drop in price and 16 fold drop in weight (please note this also corresponds to a 16 fold drop in the system preheating time and condensate amount).

Another plumbing option is thermoplastic piping, – no rust, easily glued, quickly assembled, and leak-proof. An Aquatherm polypropylene piping system operates at temperatures up to 200°F (93°C) at 15–100 psig (1–6.8 atm), so technically it should handle saturated water vapor at 6"Hg/200°F (93°C) without any problem. Like Propress, Aquatherm is backed by a 50 year warranty, and available at larger diameters (up to 300-600 mm). Polysulfone thermoplastic is already approved for low pressure steam systems in the US – up to 230°F at 15psig, it can be welded, sawed, and glued like Polyvinyl Chloride plumbing.

Clamped silicone/rubber fittings can be utilized for plastic piping; additional benefits include easy assembly/disassembly/repairs/adjustment, compensation for thermal expansion/contraction, and noise reduction. A polysulfone tube with clamped silicon fittings worked in a tested vacuum heating system for 3 winters without any problem. Liquid Crystal Polymers with heat deflection temperature of 290C can be utilized at higher temperatures.

Modern methods of leak detection allow annual proactive testing to find new leaks and keep the system operational for many years. Needless to say, thermoplastic piping with no rust/oxidation problems is of low maintenance and easy to repair.

6.2 Cast aluminum radiators

In steam heating systems pipes accounted for 30–35% of the total system weight and have to be reheated to 214°F every heating cycle. To reduce the heat loss of pipe reheating, heavy radiators are employed to accumulate heat. This archetype can be dismissed if much lighter copper or polymer piping is utilized.

Modern cast aluminum radiators present a much less expensive, lightweight alternative, they are slim, modern looking, have a modular design (length adjustable) and warrantied for up to 20 years at a pressure 7 atm [SIRA]. Such radiators by SIRA have been working problem free in a tested vacuum heating system since 2015 [26].

6.3 New control paradigm

Presently, vacuum heating systems control heat distribution via control valves on the supply lines. These are expensive items because of steam rating and large pipes diameters. Instead, the new technology employs control valves on smaller diameters vacuum lines in order to direct steam into the required system partitions. Normally open valves close when the vacuum line temperature rises above 30-40°C and are never exposed to hot vapor so a long life span is expected. Closed valves prevent air removal from a particular radiator(s), and therefore result in only partial heating (similar to TRV operation). Heat distribution and a sequence of heat supply into system partitions can be dynamically controlled.

Instead of radiators, air handlers on every floor and heat exchangers for distributed hydronic subsystems can be employed in new installations. Compared to hydronic only system, a significant size reduction of air handlers and heat exchangers can be achieved due to higher steam specific heat capacity, linear velocity and better heat transfer. Vacuum heating system plumbing for such "hybrid" systems would be simplified and heat distribution would be controlled easier. An additional benefit would be the possibility of per floor/apartment heat consumption metering in distributed subsystems, which is problematic in steam/vacuum systems.

6.4 Cogeneration heat and power (CHP) benefits and geothermal outlook

Steam/vacuum heating can be readily integrated into CHP, the most reliable and efficient electricity and heat source. In high efficiency power plants, a steam turbine exit is connected to a condenser to extract maximum electricity. The vacuum in a condenser is created via steam condensation by cold water. If the building vacuum system is connected instead of a condenser, this additional electricity would be received without steam and cold water expenditures, cooling tower, pumps, etc. In summer time, any excess steam of high temperature can be utilized in an adsorption cooler on the roof (more efficiently than hot water). Again, no electricity wasted on pumping water.

Steam is a by-product of untapped green technology - geothermal heat. "2000 times US annual energy use could be supplied indefinitely 24/7 using existing

Engineered Geothermal Systems (EGS) and perhaps 10 times as much with improved technology" [27]. At 5.5 km depth, 175-225°C heat source can be reached on the most of the US west part, and 100-150°C on the east part. A mature EGS can supply enough electricity for 800 to 41,000 average U.S. homes or dozens high-risers. The $5-10 M cost of drilling (4 and 6 km deep well, correspondingly) [28] is not an outrageous expense for reliable heat and power source compared to $7-20 M average floor price tag for high-rise in Chicago and NYC [29]. At 500% efficiency [30] this opportunity dwarfs the promises and attractiveness of HP, wind and solar power.

7. Conclusion

The US is not alone in dealing with steam heating upgrades, steam is also widespread in Europe [8] and China [31]. Worldwide there are thousands of steam heated buildings, including high-risers, which would benefit from a simple, efficient and reasonably inexpensive conversion into vacuum heating. Modern plumbing technology and materials makes vacuum heating a very attractive choice for new installations as well.

It's not unusual that old technology gets a second chance thanks to the progress in knowledge and materials. In 1893, electric cars lost the Paris-Rouen race and they have made a magnificent comeback since then. Who is to say that will not happen to vacuum heating technology for new buildings, especially skyscrapers?

Acknowledgements

The research on NextGen vacuum heating was inspired by Dan Holohan books and advice, supported by Kith Miller of NGRID and Ed Infantino of A&M SERVICES.

Abbreviations

CHP	Heat and power cogeneration
EGS	Engineered Geothermal Systems
HP	heat pump
HWH	hot water heating
LEED	Leadership in Energy and Environmental Design
MSHP	Multi Split Heat Pump
TRV	Temperature Regulating Valve
PRV	pressure regulating valve
VRF	Variable refrigerant flow
Units1SF	0.0929 M2
1Psi	0.068046 ATM
1"Hg	0.033421 atm
1 mph	0.44704 m/s
1 inch	2.54 cm
Mile	1.60934 kilometer

Author details

Igor Zhadanovsky
Applied Engineering Consulting, Newton, MA, USA

*Address all correspondence to: izhadano@gmail.com

References

[1] The Skyscraper Center data for US. Available from: http://www.skyscrapercenter.com/compare-data/submit?type%5B%5D=building&status%5B%5D=COM&status%5B%5D=UC&status%5B%5D=UCT&status%5B%5D=OH&status%5B%5D=NC&status%5B%5D=PRO&status%5B%5D=VIS&status%5B%5D=DEM&base_height_range=5&base_company=All&base_min_year=1885&base_max_year=9999&skip_comparison=on&output%5B%5D=list

[2] Pietrzak J. Development of high-rise buildings in Europe in the 20th and 21st centuries. 2014. Available from: https://yadda.icm.edu.pl/baztech/element/bwmeta1.element.baztech-07ec0c2b-8817-4191-bde1-0e28e7a1361b/c/chmot54_07.pdf

[3] Tobias M. Which HVAC configuration offers the highest efficiency for tall buildings. 2019. Available from: https://global.ctbuh.org/resources/papers/download/4206-which-hvac-configuration-offers-the-highest-efficiency-for-tall-buildings.pdf

[4] Hamlin L. Germany's Energiewende program exposed as a catastrophic failure. 2019. Available from: https://wattsupwiththat.com/2018/09/30/germanys-energiewende-program-exposed-as-a-catastrophic-failure/

[5] Tompkins G. Why Steam? An ABMA White Paper. 2020. Available from: https://www.hpac.com/association-solutions/article/21120644/boiler-systems-engineering-steam

[6] Shapiro I. Water & Energy Use in Steam-Heated Buildings. 2010. Available from: https://www.taitem.com/wp-content/uploads/SteamBoilerReplacements.pdf

[7] Urban Green Council. DEMYSTIFYING STEAM. 2019. Available from: https://www.urbangreencouncil.org/content/projects/demystifying-steam-report

[8] Gallo E. Skyscrapers and District Heating, an inter-related History 1876-1933. 2005. Available from: https://halshs.archives-ouvertes.fr/halshs-00003873v2/file/churbGC.pdf

[9] Choi J, Ludwig P. Steam System Balancing and Tuning for Multifamily Residential Build. 2012. Available from: https://www1.eere.energy.gov/buildings/publications/pdfs/building_america/steamsys_balance.pdf

[10] PARR. Steam System Balancing and Tuning for Multifamily Residential Buildings. 2011. Available from: https://www.energy.gov/sites/prod/files/2013/11/f5/steam_system_balancing_tuning_multifamily_residential.pdf

[11] Oland C. Review of Orifice Plate Steam Traps. 2001. Available from: https://www.energy.gov/sites/prod/files/2014/05/f15/orificetraps.pdf

[12] Bobker M, Kinsler ER. Balancing apartment building heating with thermostatic radiator valves. 1995. Available from: https://www.armstronginternational.com/sites/default/files/resources/documents/balancingapartmentbuildingheatingwiththermostaticradiatorvalves9-95.pdf

[13] Radiator Lab. 2014. Available from: https://www.nyserda.ny.gov/Partners-and-Investors/Clean-Energy-Startups/NYSERDA-Catalyzes-Investments/2014-09-04-Radiator-Labs

[14] Dentz J, Podorson D. Mini-Split Heat Pumps Multifamily Retrofit Feasibility

Study. 2014. Available from: https://www.nrel.gov/docs/fy14osti/61620.pdf

[15] Rieber. 179 Henry Street A Case Study in Converting from Two-Pipe Steam to Hydronic Heating. 2012. Available from: https://homeenergy.org/show/article/nav/heating/id/1806

[16] Tobias M. Steam Systems in Chicago are Outdated but Still in Use. 2020. Available from: https://www.ny-engineers.com/blog/steam-systems-in-chicago-are-outdated-but-still-in-use

[17] Al-Kodmany K. Green Retrofitting Skyscrapers: A Review. 2014. Available from: https://www.researchgate.net/publication/307669949_Green_Retrofitting_Skyscrapers_A_Review

[18] Christopher D. 200 years of LEED. 2013. Available from: https://www.usgbc.org/articles/200-years-leed-or-20-historic-buildings-you-probably-didn%E2%80%99t-know-were-green

[19] DOE. The Greening of the White House. 1999. Available from: https://clintonwhitehouse5.archives.gov/media/pdf/greening.pdf

[20] Tangherlini D. At Treasury, Green is Our Favorite Color – But We'll Take (LEED) Gold!. 2011. Available from: https://obamawhitehouse.archives.gov/blog/2011/12/21/treasury-green-our-favorite-color-well-take-leed-gold

[21] Holohan D. The Lost Art of Steam Heating. Bethpage, NY; 2004. p. 251. Available from: HeatingHelp.com

[22] Clark L. Save That Older STEAM SYSTEM. 2003. Available from: https://d1qkyo3pi1c9bx.cloudfront.net/D7021408-671F-42FA-837E-DEB20A6B3D76/b9c37f57-bdda-4bf5-8467-3fb5a7df0ab2.pdf

[23] Holohan D. A New Look at Vacuum Heating. P&M Magazine; 2015. Available from: https://www.pmmag.com/

[24] US Patent 9027846. Vacuum sustaining heating systems and methods. Available from: https://www.freepatentsonline.com/9027846.html

[25] NGRID. Technical Assistance Study. Vacuum Steam Heating. 2014. Available from: https://drive.google.com/open?id=0B-5hKw5FhaA4am1Eb2VyaUNwR3c%20

[26] Zhadanovsky I. Vacuum Steam Heating: Time for a Comeback? 2017. Available from: https://www.facilitiesnet.com/hvac/contributed.aspx?id=38418

[27] Chandler D. Power from down under. 2010. Available from: https://www.usgbc.org/articles/200-years-leed-or-20-historic-buildings-you-probably-didn%E2%80%99t-know-were-green

[28] Tester J. Geothermal direct use. 2015. p. 50. Available from: https://www.energy.gov/sites/prod/files/2015/09/f26/Geothermal%20Direct%20Use%20Workshop%20Summary%20Report%20-%20No%20List%20%2809-23-2015%29.pdf

[29] Barr J. The Economics of Skyscraper Height (Part IV). 2019. Available from: https://buildingtheskyline.org/skyscraper-height-iv/

[30] Egg J. Geothermal Marketplace in the Eastern U.S. 2015. p. 88. Available from: https://www.energy.gov/sites/prod/files/2015/09/f26/Geothermal%20Direct%20Use%20Workshop%20Summary%20Report%20-%20No%20List%20%2809-23-2015%29.pdf

[31] Xia J. Analysis of current status and energy saving of steam system. 2009. Available from: https://www.lunwen163.com/h/152/155/6830.html

Chapter 5

Advancements in Indirect Evaporative Cooling Systems through Novel Operational Configuration

Muhammad Ahmad Jamil, Muhammad Wakil Shahzad, Ben Bin Xu, Muhammad Waqar Ashraf, Kim Choon Ng, Nida Imtiaz and Haseeb Yaqoob

Abstract

Rising global temperature has triggered the cooling demand in the last three decades with growing predictions for the future. The use of conventional energy-intensive and high global warming chemical-based cooling systems is working in a loop, increasing the global warming rate, emissions, and cooling system inventory. Therefore, the development of an innovative cooling system with high energy efficiency, low monetary cost, and environmentally sustainable. The indirect evaporative cooling-based systems have shown potential to serve the purpose because of low energy consumption, absence of energy, and cost-intensive equipment like compressors and water-based operation. A novel indirect evaporative cooler based on an innovative operational configuration is proposed, fabricated, and tested experimentally. The Proposed system has several advancements compared to the conventional indirect evaporative coolers like high operational reliability, low maintenance, and better control of the processes in the system. The study shows that the proposed system can achieve a temperature drop of as high as 14°C. The maximum cooling capacity of the system is calculated as 110 W, and the cooling performance index of 28. The performance of the cooler improves with increasing outdoor air temperature which makes it suitable for diverse climatic conditions. Moreover, the proposed design offers several benefits due to novel operational configurations by addressing limitations in the earlier systems.

Keywords: advanced evaporative cooler, humidity-controlled cooling, cleaner air conditioning, sustainable development, cooling systems

1. Introduction

A remarkable surge in the global energy demand has been seen in the recent past because of an exponential rise in population, urbanization, and economic growth [1].

Moreover, the improving lifestyle is also contributing significantly to the energy consumption to maintain human comfort [2, 3]. The worldwide energy demand is estimated to rise to 850 quadrillions Btu by 2050 from merely 500 quadrillions Btu in 2010 indicating around a 50% rise in 4 decades [4]. Particularly, the situation is alarming for developing countries that will face a tremendous surge of around 71%in the energy demand compared to the developed countries with 18% [5, 6]. Building energy consumption is of critical importance in the overall energy consumption scenario because of the major share and diverse necessary activities [7]. The common among these are human thermal comfort, cleaning, cooking, food preservation, lightening, etc. Despite the wide range of activities taking place in buildings, the overall energy consumption can be classified into 6 different types as shown in **Figure 1** [8, 9]. For commercial and residential buildings, the energy consumption is as follows: heat ventilation and air conditioning 36–40%, lighting 12–20% hot water supply 9–13%, electronics 8–15%, refrigeration 5–7%, cooking 4–5% and others 8–18% [10, 11].

Meanwhile, the demand for air conditioning is also increasing continuously and the global air conditioner inventory is expected to cross 5600 million by 2050 from merely 1600 million units reported in 2016 [12]. The corresponding energy consumption and emission are also expected to surge to 6200-Terawatt hour and 170 Gigatons by 2050 [13]. The use of conventional vapor compression chillers is one of the major reasons for these high energy demands. This is because these systems have very low energy efficiency, which is low efficiency, and involve high global warming potential refrigerants for cooling [14]. Meanwhile, their performance has not seen any considerable improvement in the last 30 years with a stagnant coefficient of performance (COP) of 3–4. This is because of rigid temperature lifts (5 to 7°C) across the evaporator and condenser which require a high input energy compressor [15]. Moreover, high global warming potential chemical refrigerants are used for compression and expansion in the system for cooling with high chances of leakage due to elevated pressure operation. These problems cannot be abandoned in conventional operational schemes. So, an innovative system is required to achieve a breakthrough in the cooling sector [16].

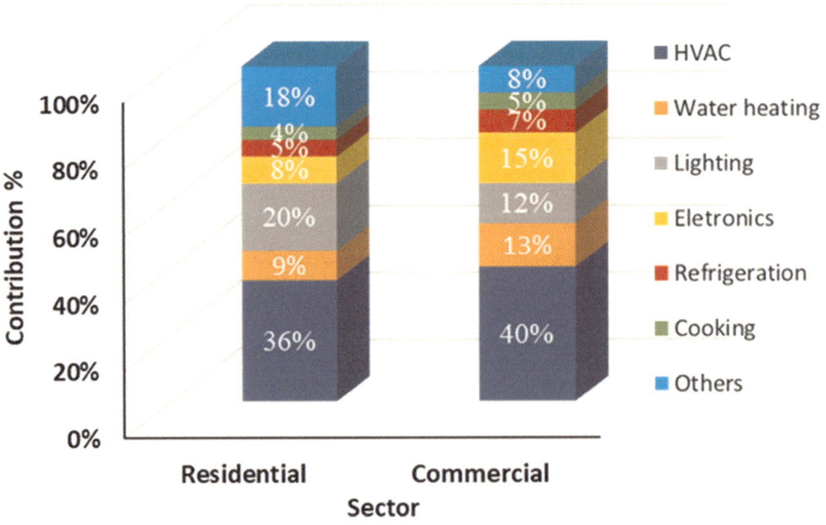

Figure 1.
Building energy consumption distribution [8–10].

One of the lucrative options for the above-mentioned problems emerged is the indirect evaporative cooling system [17]. It uses a water-based cooling mechanism and does not involve any hazardous chemical refrigerant [18]. Moreover, it does not involve any high energy consumption compressor. Rather these systems use fans for air movement through the system and pump for water supply [19, 20]. In these systems, the hot outer air is cooled using cold wet air (water-air mixture) in two different channels. The channels are separated by thermally conductive impermeable walls which only allow heat exchange between the air streams without any moisture exchange. Therefore, these systems produce cold dry air using air and water through evaporation [21, 22].

These systems have been extensively studied by the research community from an experimental and theoretical viewpoint. For instance, multipoint air injection in IECs improved the cooling performance by achieving an additional 2 to 3°C temperature drop by achieving COP as high as 78 for cooling [23]. Likewise, the enhancement of heat transfer plates through protrusion was reported to enhance the IEC performance achieving wet bulb efficiency up to 85% [24, 25]. Chua et al. [26] investigated a felt-assisted IEC with a cross-flow heat exchanger and showed that the cooler can achieve wet bulb efficiency of 90%. Similarly, Duan et al. [27] developed a compact heat exchangers based counter flow IEC. They reported the performance characteristics of the system in terms of wet bulb efficiency up to 107%, cooling capacity up to 8.5 kW, and energy efficiency ratio up to 20. They also improved the system efficiency and energy efficiency ratio by 30% and 40%, respectively by enhancing plates with corrugations [28]. Similarly, finned channel-based systems have also shown promising performance with wet bulb efficiency as118–122% and dewpoint effectiveness of 75–90% [29]. Cui et al. [30] investigated the hybrid air conditioning system with IEC and reported the COP as 14.2 for cooling.

Besides experimental studies, the theoretical analyses of IEC systems have also shown considerable performance improvements through design and operational modifications. For instance, Oh et al. [31] studied the effect of purge configuration on regenerative type IEC performance. They reported the maximum cooling performance at 35% purge ratio with a dew point efficiency of 58% and cooling capacity of 59 W. Similarly, Pandelidis et al. [32] showed that the performance of regenerative IEC systems can be improved through additional perforations, particularly at higher air flow rate ratio > 45%. Riangvilaikul et al. [33] showed that the Polyurethane based IEC system showed promising performance by achieving dewpoint efficiency of 65–87% and the wet bulb efficiency 106–109%. Wang et al. [34] optimized a dewpoint regenerative type IEC based on the entropy principle. The optimal values for velocity, heat exchanger length, channel gap, and airflow rate ratio were reported as 1.0 m/s, 1–1.75 m, 3-5 mm, and 30–40%, respectively. Jradi et al. [35] optimized the dew point cooler for maximum wet bulb efficiency of 112%, and dewpoint efficiency of 78%. The optimal channel gap and heat exchanger length under considered operating conditions were reported as 5 mm and 500 mm, respectively. Similarly, Adam et al. [36] optimized the cross-flow IEC and reported the supply air temperature as 24°C and wet bulb efficiency as 92%. and $T_{PA,o}$ = 24°C. Some other efforts include the study of the effects of condensation in the dry channel [37], wettability enhancement in the wet channel [38], and improving heat transfer characteristics of heat exchangers [39].

The literature review shows that indirect evaporative coolers are a capable substitute for compression cooling systems because of promising performance from temperature drop, efficiency, and COP. Therefore, an innovative indirect evaporative cooler with a novel operational configuration is proposed and tested experimentally.

The proposed system has several advancements like simple construction, separate control for the air-water mixing process, fewer maintenance requirements, and better control of sensible and latent heat transfer processes. The current study analyzes a generic cell of the system which can be used to develop the design metrics for commercial-scale expansion of the proposed idea.

2. Proposed system

2.1 Design concept

The current system consists of the following major components: a heat exchanger (HE), a mixing chamber (MX), fans, a water pump, and nozzles as shown in **Figure 2**. The heat exchanger consists of three identical channels designated for supply air (1 middle channel) and working air (2 side channels). The channels are constructed using a conductive impermeable wall. Spacers are used to maintain the required channel gap to avoid sheet bulging at high air flow rates. The mixing chamber is constructed using acrylic sheets. The chamber is equipped with water spray systems (nozzles, pump, and supplementary water float inlet), and working air manifolds. The sump of the humidifier contains water that is used in a recirculation manner for cooling the working air. As the water is consumed during evaporation, the supplementary water is fed to the system from the reservoir through a float-controlled valve. Besides, the system is also equipped with two axial fans for product air and working air supply. The working cycle of the system is presented below.

In the proposed system the hot outer air undergoes sensible cooling (1 to 2) in the middle (dry) channel of the heat exchanger. The outdoor air flows from the bottom of the channel to the top. An axial fan is employed to maintain the required flow rate and overcome the channel frictional pressure drop. This air stream rejects the heat to the working air in the two side channels flowing in the counter-current direction (from top to bottom). While the working air stream first enters the mixing chamber (as hot outer air) where it is mixed with fine water mist generated using atomizing nozzles. This air-water meshing causes the hot air to release its heat through evaporation and gain high humidity (100%) (1 to 3). This cold humid working air carrying fine water particles enters the heat exchanger and extracts heat from the dry channels. The humidity drops and the temperature rises as the working air moves along the heat exchanger because of latent and sensible heat transfer respectively (3 to 4). The product air (dry and cold) is supplied to the conditioned space and the working air (hot and humid) is discarded to the outer environment.

2.2 Experimental test rig

Based on the above-presented design concept, an experimental test rig was developed as shown in **Figure 3**. Different components of the system are labeled as, A: heat exchanger, B: mixing chamber, C: outer air inlet to heat exchanger, D: product air outlet, E: air inlet to MX, F: working air inlet to HE, G: working air outlet, and H: atomizing nozzles. The heat exchanger was constructed using 0.025 mm thick high thermal conductivity (235 W/m K) Aluminum sheets. The channel gap was maintained at 5 mm using closed foam spaces. The acrylic sheets were used to support the heat exchanger structure and act as adiabatic walls for working air channels. The

DOI: http://dx.doi.org/10.5772/intechopen.107305

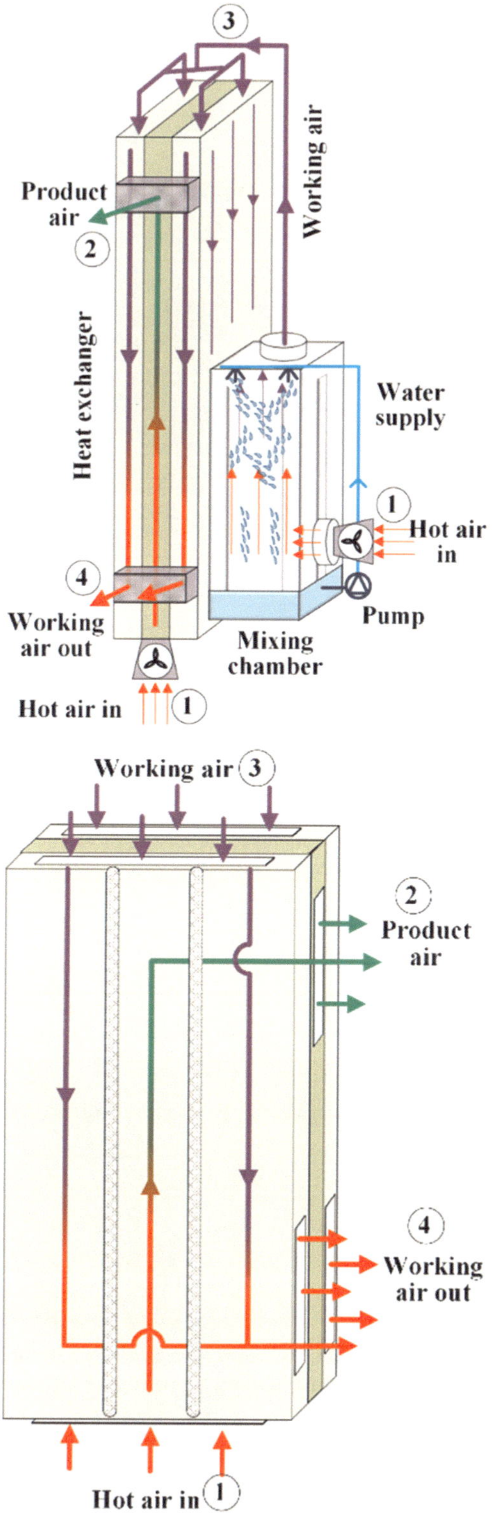

Figure 2.
Schematic diagram of the proposed system.

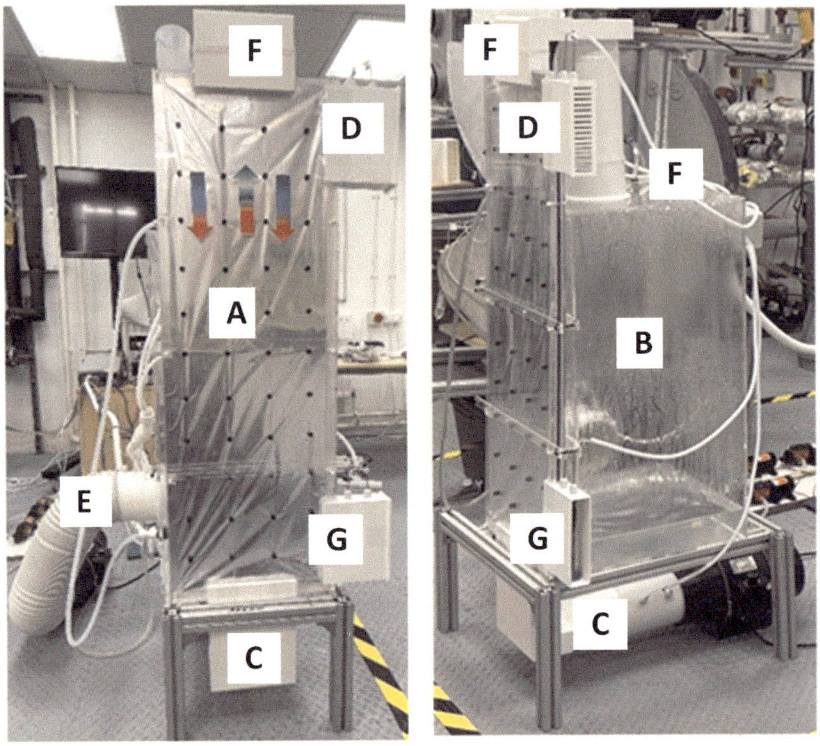

Figure 3.
Experimental test rig front and side view.

Parameter	Value
Heat exchanger length, mm	1000
Heat exchanger width, mm	300
Heat exchanger wall thickness, mm	0.025
Channel gap, mm	5
Mixing chamber height, mm	800
Mixing chamber width, mm	400
Mixing chamber length, mm	400

Table 1.
Geometric characteristics of the system.

inlet and outlet manifolds of the heat exchanger were developed using an in-house 3D printing facility. The design characteristics of the system are summarized in **Table 1**.

A real-time data acquisition system is used to monitor and record temperature data during experimentation. For this purpose, dry and wet bulb temperature sensors are installed at different locations in the system. General purpose thermistor probes are used for temperature measurements. For wet bulb temperature, the measuring station is developed using high capillary action felt material with a continuous water supply. Meanwhile, the flow rates in the system are measured using a hot wire anemometer (Testo 405i) at product air and working air outlets.

Parameter	Value
Outer air temperature, °C	29–43
Outdoor air humidity, g/kg	10
Outer air velocity, m/s	2.5
Working air velocity, m/s	2.5
Working air humidity, g/kg	18

Table 2.
Process parameters.

Parameter	Description	Accuracy
Temperature	General purpose thermistor probes (by OMEGA)	±0.15°C
Velocity	Hot wire anemometer (by Testo 405i)	±0.1 m/s
Humidity	Wet bulb measuring station (by OMEGA and customized)	±0.15°C
Data logging	Agilent Benchlink 34790a	

Table 3.
Instrumentation details.

Detailed experimentation of the developed system was conducted to capture the performance picture of the system. For this purpose, the system is operated at varying outdoor air temperature conditions. The process parameters considered in the study are presented in **Table 2**. The instrumentation details are summarized in **Table 3**.

2.3 Performance assessment

The performance of the cooler is measured in terms of heat extraction (cooling) capacity from the product air stream as it moves along the heat exchanger from 1 to 2. It is calculated in terms of airflow rate, specific heat, and temperature drop [40, 41].

$$\dot{Q}_{1-2} = \dot{m}_{PA}\, c_p\, (T_1 - T_2) \tag{1}$$

The heat rejected by the dry channel air stream during processes 1–2 is taken by the working air stream in the wet channel (3–4) as a combined latent and sensible heat. This heat transfer process is given as:

$$\dot{Q}_{1-2} = \dot{Q}_{3-4} = \dot{m}_{WA}\, c_p\, (T_4 - T_3) + \dot{m}_{vapor}\, \lambda_{fg} \tag{2}$$

The cooler performance is also measured in terms of the cooler performance indicator (CPI) which is the ratio of cooling produced to the energy consumed [42].

$$CPI = \frac{\dot{Q}_{1-2}}{\dot{E}_{input}} \tag{3}$$

The energy consumption is calculated in terms of fan energy and pumps energy used to maintain the required air and water flow rates, respectively. It can be used in terms of fluid flow rate, pressure differential, and component efficiency [43].

$$\dot{E}_{in} = \dot{E}_{blower} + \dot{E}_{pump} = \frac{\dot{V}_{air}\,\Delta P_{HE}}{\eta_{blower}} + \frac{\dot{V}_{water}\,\Delta P_{water}}{\eta_{pump}} \tag{4}$$

The pressure drop in the heat exchanger channels is calculated in terms of friction factor, channel length, flow velocity, and equivalent diameter [40, 44].

$$\Delta P_{ch} = f\frac{L}{D_h}\frac{\rho V_{air}^2}{2} \tag{5}$$

The fan and pump power are calculated using pressure drops of air and water with corresponding flow rates. The maximum power input calculated is 4 W which is constant for all operating conditions.

The above equations are valid under the following standard assumptions.

- Steady-state performance
- Negligible heat leak to-and-from the system
- The heat rejected by the product air stream is absorbed by the working air stream
- Supplementary water is fed at a constant temperature equivalent to a wet bulb during longer operational times
- Pressure drop in connections and manifold is negligible

3. Results and discussion

The performance of the proposed indirect evaporative cooler was investigated in terms of temperature drop, cooling capacity, and cooling performance index. The most important parameter in indirect evaporative coolers is the supply air temperature. This is because all other performance indices are governed by the supply air temperature. For this purpose, the cooler was tested under different outdoor air conditions to record the supply air temperature trend for 4-to-5-hour continuous operation. **Figure 4** shows the typical product air temperature trends for an outdoor air temperature of 40 ± 0.5°C. It shows that the product air was obtained at a uniform temperature of 25 ± 0.5°C. It implied the steady cooler performance during the whole operational time producing the uniform cool product air. Therefore, the cooler is suitable for continuous longer operations without the development of any longitudinal heat conduction effect or heat storage in heat exchanger walls. The visual inspection during experimentation shows the mist evaporation on the walls which extracts heat from the wall thus keeping it at the same temperature. Meanwhile, it is also important to emphasize that the product air was supplied at constant absolute humidity of 10 g/kg because of sensible cooling.

The temperature trends for the working air stream are shown in **Figure 5**. It is observed that the hot outer air entered at 40 ± 0.5°C in the mixing chamber. It was cooled to the wet bulb temperature of 23 ± 0.5°C by mixing with water mist. During

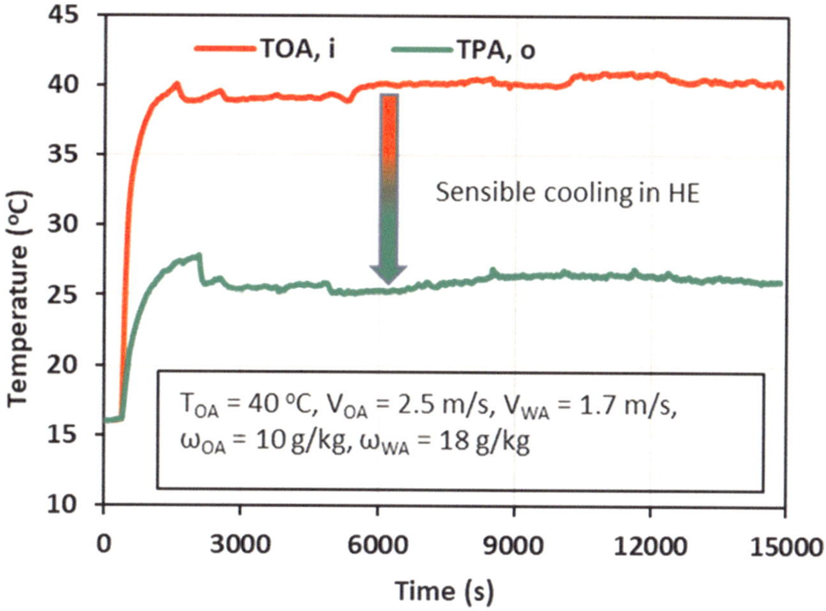

Figure 4.
Product air temperature trend.

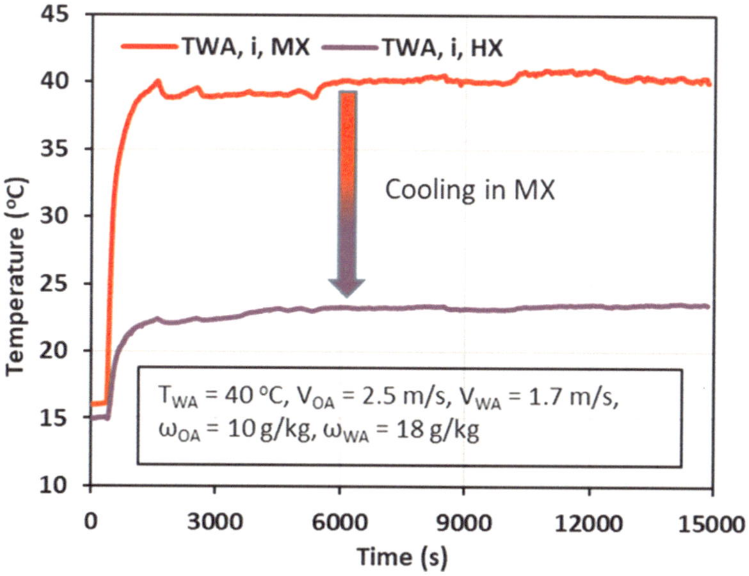

Figure 5.
Working air temperature trend.

mixing the working air achieved 100% relative humidity (ω = 18 g/kg). This cold and humid air then enters the heat exchanger and extracts heat from the product air stream. **Figure 6** shows the temperature trends for working air at the heat exchanger inlet and the product air at the heat exchanger outlet. It showed that the cooler performed close to wet bulb temperature with a maximum temperature differential across the air streams of 2–3°C thus giving the highest cooling performance.

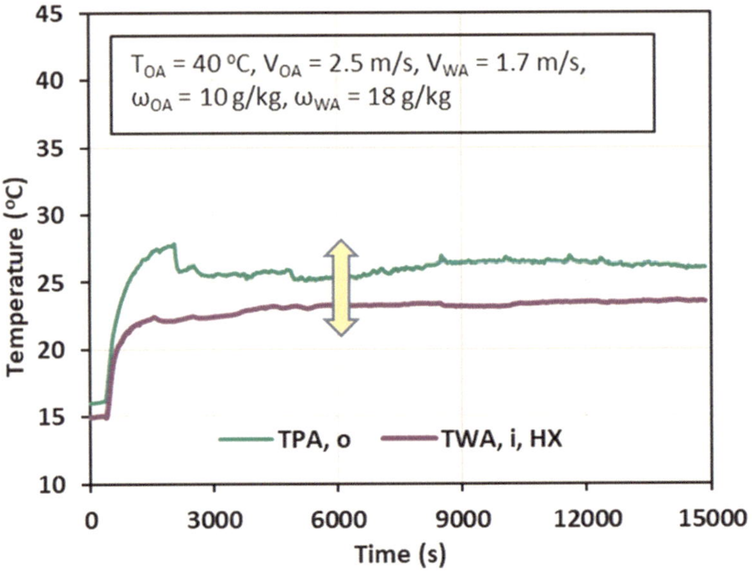

Figure 6.
Product air and working air temperature trend.

The performance of the cooler in terms of product air temperature at different outer air conditions is presented in **Figure 7**. It shows that the product air temperature increased as the outer air inlet temperature increased. A temperature rise of around 7° C (from 22.6 to 29.3°C) was observed for the outer temperature rising from 29 to 43° C. It suggested that the cooler can be used for any other higher or lower outer air temperature conditions with slight variation in the product air temperature. However, it is also important to emphasize that the variation in product air temperature is less

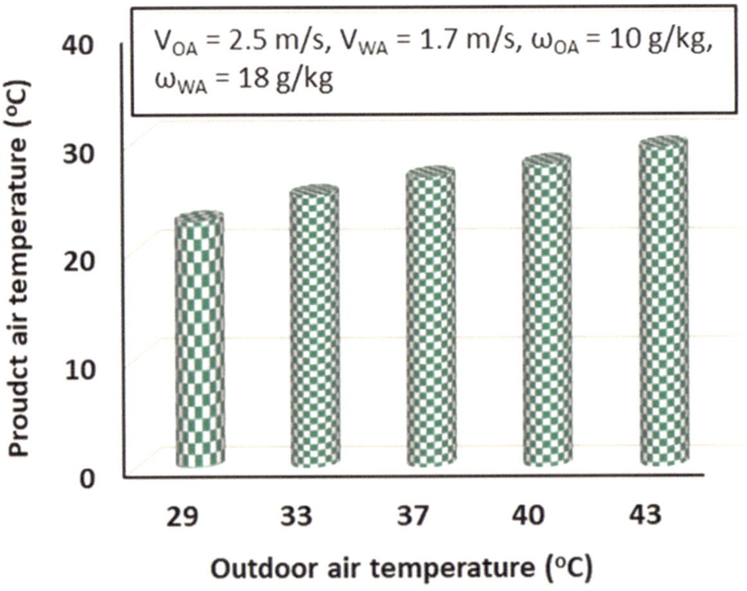

Figure 7.
Product air temperature trends at different outer air temperatures.

compared to the outer air conditions. It showed the performance of the cooler close to the wet bulb temperatures for all the operational conditions.

The effect of outer air temperature on the cooling capacity is presented in **Figure 7**. Correlated to the temperature drop, the cooling capacity of the cooler was observed to be increasing with the increasing outer air. For instance, the cooling capacity varied from 50 W to 110 W as the outer air temperature increased from 29 to 43°C. This increase in cooling capacity was achieved due to increasing temperature drop at higher outer temperature conditions. Similarly, the cooling performance index at different outer air temperature conditions is presented in **Figure 8**. It showed that the cooling performance index increased at higher outer air temperature conditions. The increase in CPI was due to an increase in the cooling capacity. For instance, the CPI increased from 13 to 28 as the outer air temperature increased from 29 to 43°C. Meanwhile, it is also worth mentioning that the power input was considered constant for all the cases because of the fixed pump and fan installation. However, less power is required (theoretically) at lower outer air conditions because of low water requirement and less working air velocity. Therefore, intelligent system control regulating the air and water supply commensurate to the outer air temperatures can offer the same higher CPI at lower outer air temperature conditions (**Figure 9**).

Besides promising energy performance, the proposed system also resolves the issue of high humidity in conventional water-based cooling systems. The outlet of the system for all operating conditions remains within the comfortable zones recommended by ASHRAE (Winter: RH = 75%, T = 21°C, Summer: RH = 53%, T = 27°C) and ISO (Winter: RH = 30–70%, T = 23°C, Summer: RH = 30–70%, T = 26°C). So, the issues associated with high humidity are eliminated while using the proposed system. One of the issues in this regard is Legionnaires which commonly occur due to airborne water droplets. It is important to emphasize that the supply air in the proposed system does not interact

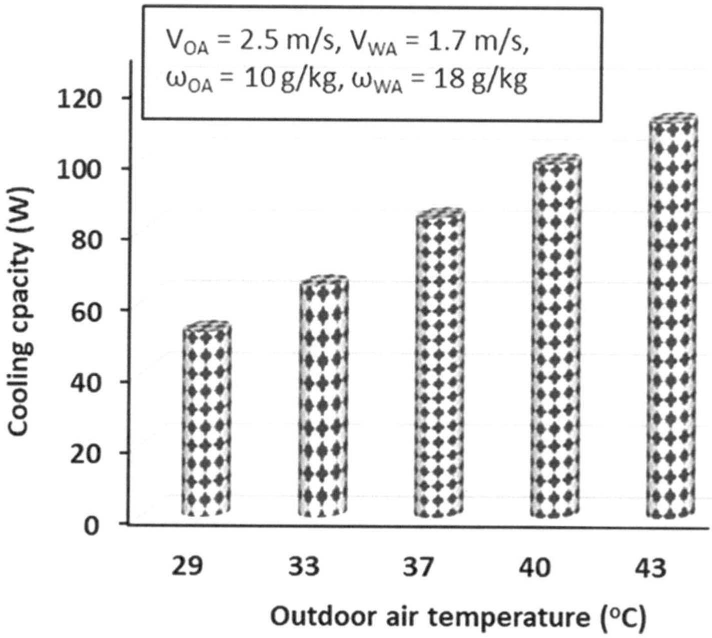

Figure 8.
Cooling capacity at different outer air temperatures.

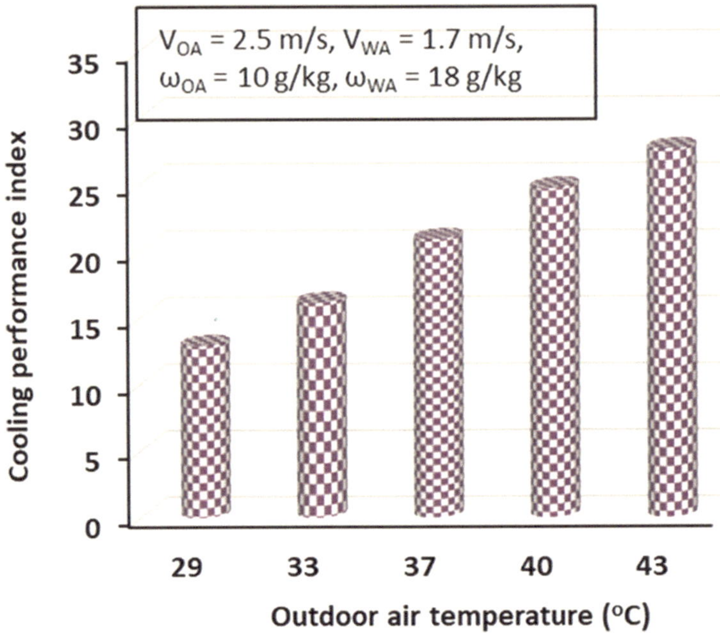

Figure 9.
Cooling performance index at different outer air temperatures.

with water during any operational stage. Therefore, there is no chance of mixing water droplets with the supply air stream and associated problems. Moreover, the working air can be used for beneficial use like hydroponics or water harvesting to further mitigate the potential issues with wet air delivery to the ambient.

The system is also convenient to integrate with the building particularly using the existing ducting network. This is because the system provides cool dry air at the required temperature as a single outlet. The outlet of the system can be connected to the ducting network in the buildings to supply air at different points. The duct fans can be used to facilitate air delivery at the required points by overcoming the pressure drop in ducts. It is also important to mention that in industrial-scale systems with high-velocity air delivery, the velocity of air is managed in the distribution duct to achieve the required comfortable range of supply air velocity as per ASHRAE standard 55.

4. Concluding remarks

An innovative indirect evaporative cooler is developed and tested experimentally. The proposed design is aimed to resolve the critical issues from manufacturing and operational viewpoints. The salient features of the proposed system are simple construction, low manufacturing, less maintenance, high operational reliability, and competitive cooling performance. The major outcomes of the study are summarized below.

- The system showed stable performance by cooling the product air to 25 ± 0.5°C from an outer air temperature of 45 ± 0.5°C.

- The product air temperature showed an increase of 7°C from 22.6 to 29.3°C as the outer air inlet temperature increased from 29 to 43°C. It suggested that the cooler can be used for any other higher of lower outer air temperature conditions with slight variation in the product air temperature.

- Correlated to the temperature drop, the cooling capacity of the cooler was observed to be increasing from 50 W to 110 W with increasing outer air from 29 to 43°C. This increase in cooling capacity was achieved due to increasing temperature drop at higher outer temperature conditions.

- The cooling performance index increased at higher outer air temperature conditions due to an increase in the cooling capacity. The CPI increased from 13 to 28 as the outer air temperature increased from 29 to 43°C.

Overall, the system showed promising performance with a maximum temperature drop of 14°C, a maximum cooling capacity of 110 W, and a maximum cooling performance index of 43°C. Meanwhile, the system supplied product air within the ASHRAE comfortable range for all the considered operational conditions. In addition to competitive performance, the system also addressed the limitations of the conventional systems through rearrangement and retrofitting of the component. Therefore, the proposed system can be a lucrative option for industrial-scale expansion of indirect evaporative coolers.

Acknowledgements

The authors would like to thank Northumbria University UK for funding under reference #RDF20/EE/MCE/SHAHZAD, and Northern Accelerator Proof-of-Concept award for AD4DCs (NACCF-232) Awarded to Dr. Muhammad Wakil Shahzad. Also acknowledged is the support provided by KAUST cooling initiative.

Conflict of interest

The authors declare no known conflict of interest.

Nomenclature

A	heat transfer area, m^2
c_p	specific heat
$\dot{E}$	energy, W
f	friction factor
$\dot{m}$	mass flow rate, kg/s
P	pressure, pa
$\dot{V}$	volume flow rate

Greek Symbols

Δ	change in quantity
λ	enthalpy

Subscripts

i	inlet
o	outlet
OA	outer air
PA	product air
WA	working air

Abbreviations

CPI	cooling performance index
IEC	indirect evaporative cooler

Author details

Muhammad Ahmad Jamil[1*], Muhammad Wakil Shahzad[1*], Ben Bin Xu[1], Muhammad Waqar Ashraf[2], Kim Choon Ng[3], Nida Imtiaz[1,4] and Haseeb Yaqoob[5]

1 Mechanical and Construction Engineering Department, Northumbria University, Newcastle Upon Tyne, UK

2 Department of Chemical Engineering, Centre for Process Systems Engineering, University College London, London, UK

3 Water Desalination and Reuse Center, King Abdullah University of Science and Technology, Thuwal, Saudi Arabia

4 Department of Mechanical Engineering, Universiti Teknologi Malaysia, Johor Bahru, Malaysia

5 Department of Automotive Engineering Technology, Punjab Tianjin University of Technology, Lahore, Pakistan

*Address all correspondence to: muhammad2.ahmad@northumbria.ac.uk and muhammad.w.shahzad@northumbria.ac.uk

References

[1] UNFCCC. The Paris Agreement, United Nations Framework Convention on Climate Change. Available from: https://unfccc.int/process-and-meetings/the-paris-agreement/the-paris-agreement. [Accessed: October 3, 2021]

[2] Paltsev S, Morris J, Kheshgi H, Herzog H. Hard-to-abate sectors: The role of industrial carbon capture and storage (CCS) in emission mitigation. Applied Energy. 2021;**300**:117322. DOI: 10.1016/J.APENERGY.2021.117322

[3] Liu B, Liu S, Xue B, Lu S, Yang Y. Formalizing an integrated decision-making model for the risk assessment of carbon capture, utilization, and storage projects: From a sustainability perspective. Applied Energy. 2021;**303**: 117624. DOI: 10.1016/J.APENERGY.2021.117624

[4] Annual Energy Outlook 2020 with projections to 2050. Available from: https://www.connaissancedesenergies.org/sites/default/files/pdf-actualites/AEO2020% 20Full%20Report.pdf

[5] Khosla R, Miranda ND, Trotter PA, Mazzone A, Renaldi R, McElroy C, et al. Cooling for sustainable development. Nature Sustainability. 2020;**4**(3): 201-208. DOI: 10.1038/s41893-020-00627-w

[6] Kynčlová P, Upadhyaya S, Nice T. Composite index as a measure on achieving sustainable development goal 9 (SDG-9) industry-related targets: The SDG-9 index. Applied Energy. 2020;**265**: 114755. DOI: 10.1016/J.APENERGY.2020.114755

[7] Ürge-Vorsatz D, Cabeza LF, Serrano S, Barreneche C, Petrichenko K. Heating and cooling energy trends and drivers in buildings. Renewable and Sustainable Energy Reviews. 2015;**41**: 85-98. DOI: 10.1016/j.rser.2014.08.039

[8] Real Prospects for Energy Efficiency in the United States | The National Academies Press. Available from: https://www.nap.edu/catalog/12621/real-prospects-for-energy-efficiency-in-the-united-states. [Accessed January 23, 2021]

[9] Air Conditioning Use Emerges as One of the Key Drivers of Global Electricity-Demand Growth - News - IEA. Available from: https://www.iea.org/news/air-conditioning-use-emerges-as-one-of-the-key-drivers-of-global-electricity-demand-growth. [Accessed January 23, 2021]

[10] Chua KJ, Islam MR, Ng KC, Shahzad MW. Advances in air conditioning technologies. In: Green Energy and Technology. Springer Singapore: Singapore; 2021. DOI: 10.1007/978-981-15-8477-0

[11] Ikhlaq M, Al-Abdeli YM, Khiadani M. Methodology for spatially resolved transient convection processes using infrared thermography. Experimental Heat Transfer. 2021;**34**(3):269-292. DOI: 10.1080/08916152.2020.1749189

[12] ACHR. Global A/C Market Starting to Warm Up|2014-08-18|. Available from: https://www.achrnews.com/articles/127385-global-ac-market-starting-to-warm-up [Accessed July 17, 2022]

[13] Jamil MA, Xu BB, Dala L, Sultan M, Jie L, Shahzad MW. Experimental and normalized sensitivity based numerical analyses of a novel humidifier-assisted highly efficient indirect evaporative cooler. International Communications in Heat and Mass Transfer. 2021;**125**:105327. DOI: 10.1016/j.icheatmasstransfer.2021.105327

[14] Shahzad MW, Lin J, Xu BB, Dala L, Chen Q, Burhan M, et al. A

spatiotemporal indirect evaporative cooler enabled by transiently interceding water mist. Energy. 2021;**217**:119352. DOI: 10.1016/j.energy.2020.119352

[15] Cui X, Chua KJ, Yang WM. Use of indirect evaporative cooling as pre-cooling unit in humid tropical climate: An energy saving technique. Energy Procedia. 2014;**61**:176-179. DOI: 10.1016/j.egypro.2014.11.933

[16] Chen Q, Burhan M, Shahzard MW, Alrowais R, Ybyraiymkul D, Akhtar FH, et al. A Novel Low-Temperature Thermal Desalination Technology Using Direct-Contact Spray Method. In Desalination - Challenges and Opportunities. London, UK: IntechOpen; 2020. DOI: 10.5772/INTECHOPEN.92416

[17] Lv J, Meng D, Chen Y, You Y, Li H. An experimental study on condensate film of indirect evaporative cooler. Energy Procedia. 2018;**2019**(158):5753-5758

[18] Meng D, Lv J, Chen Y, Li H, Ma X. Visualized experimental investigation on cross-flow indirect evaporative cooler with condensation. Applied Thermal Engineering. 2018;**145**(September):165-173. DOI: 10.1016/j.applthermaleng.2018.09.026

[19] Sun T, Huang X, Chen Y, Zhang H. Experimental investigation of water spraying in an indirect evaporative cooler from nozzle type and spray strategy perspectives. Energy and Buildings. 2020;**214**:109871

[20] Wang F, Sun T, Huang X, Chen Y, Yang H. Experimental research on a novel porous ceramic tube type indirect evaporative cooler. Applied Thermal Engineering. 2017;**125**:1191-1199. DOI: 10.1016/j.applthermaleng.2017.07.111

[21] Chen Y, Yang H, Luo Y. Parameter sensitivity analysis of indirect evaporative cooler (IEC) with condensation from primary air. Energy Procedia. 2016;**88**:498-504. DOI: 10.1016/j.egypro.2016.06.069

[22] Chen Y, Yang H, Luo Y. Indirect evaporative cooler considering condensation from primary air: Model development and parameter analysis. Building and Environment. 2016;**95**:330-345. DOI: 10.1016/j.buildenv.2015.09.030

[23] Shahzad MW, Burhan M, Ybyraiymkul D, Oh SJ, Choon K. An improved indirect evaporative cooler experimental investigation. Applied Energy. 2019;**256**(September):113934

[24] De Antonellis S, Joppolo CM, Liberati P, Milani S, Romano F. Modeling and experimental study of an indirect evaporative cooler. Energy and Buildings. 2017;**142**:147-157. DOI: 10.1016/j.enbuild.2017.02.057

[25] De Antonellis S, Joppolo CM, Liberati P, Milani S, Molinaroli L. Experimental analysis of a cross flow indirect evaporative cooling system. Energy and Buildings. 2016;**121**:130-138. DOI: 10.1016/j.enbuild.2016.03.076

[26] Chua KJ, Xu J, Cui X, Ng KC, Islam MR. Numerical heat and mass transfer analysis of a cross - flow indirect evaporative cooler with plates and flat tubes. Heat and Mass Transfer. 2015;**52**:1765-1777. DOI: 10.1007/s00231-015-1696-y

[27] Duan Z, Zhao X, Li J. Design, fabrication and performance evaluation of a compact regenerative evaporative cooler: Towards low energy cooling for buildings. Energy. 2017;**140**:506-519. DOI: 10.1016/j.energy.2017.08.110

[28] Duan Z, Zhan C, Zhao X, Dong X. Experimental study of a counter-flow

regenerative evaporative cooler. Building and Environment. 2016;**104**: 47-58. DOI: 10.1016/j.buildenv.2016.04.029

[29] Lee J, Lee DY. Experimental study of a counter flow regenerative evaporative cooler with finned channels. International Journal of Heat and Mass Transfer. 2013;**65**:173-179. DOI: 10.1016/j.ijheatmasstransfer.2013.05.069

[30] Cui X, Islam MR, Chua KJ. An experimental and analytical study of a hybrid air-conditioning system in buildings residing in tropics. Energy and Buildings. 2019;**201**:216-226. DOI: 10.1016/j.enbuild.2019.06.028

[31] Oh SJ, Shahzad MW, Burhan M, Chun W, Kian Jon C, KumJa M, et al. Approaches to energy efficiency in air conditioning: A comparative study on purge configurations for indirect evaporative cooling. Energy. 2019;**168**: 505-515. DOI: 10.1016/j.energy.2018.11.077

[32] Pandelidis D, Anisimov S, Worek WM. Comparison study of the counter-flow regenerative evaporative heat exchangers with numerical methods. Applied Thermal Engineering. 2015;**84**:211-224. DOI: 10.1016/j.applthermaleng.2015.03.058

[33] Riangvilaikul B, Kumar S. Numerical study of a novel dew point evaporative cooling system. Energy and Buildings. 2010;**42**(11):2241-2250. DOI: 10.1016/j.enbuild.2010.07.020

[34] Wang L, Zhan C, Zhang J, Zhao X. Optimization of the counter-flow heat and mass exchanger for M-cycle indirect evaporative cooling assisted with entropy analysis. Energy. 2019;**171**:1206-1216. DOI: 10.1016/j.energy.2019.01.099

[35] Jradi M, Riffat S. Experimental and numerical investigation of a dew-point cooling system for thermal comfort in buildings. Applied Energy. 2014;**132**: 524-535. DOI: 10.1016/J.APENERGY.2014.07.040

[36] Adam A, Han D, He W, Chen J. Numerical analysis of cross-flow plate type indirect evaporative cooler: Modeling and parametric analysis. Applied Thermal Engineering. 2020; **2021**(185):116379. DOI: 10.1016/j.applthermaleng.2020.116379

[37] Min Y, Chen Y, Yang H. Numerical study on indirect evaporative coolers considering condensation: A thorough comparison between cross flow and counter flow. International Journal of Heat and Mass Transfer. 2019;**131**: 472-486. DOI: 10.1016/j.ijheatmasstransfer.2018.11.082

[38] Guo XC, Zhao TS. A parametric study of an indirect evaporative air cooler. International Communications in Heat and Mass Transfer. 1998;**25**(2): 217-226. DOI: 10.1016/S0735-1933(98)00008-6

[39] Lin J, Thuan D, Wang R, Jon K. On the fundamental heat and mass transfer analysis of the counter-flow dew point evaporative cooler. Applied Energy. 2018;**217**(February):126-142. DOI: 10.1016/j.apenergy.2018.02.120

[40] Kashyap S, Sarkar J, Kumar A. Comparative performance analysis of different novel regenerative evaporative cooling device topologies. Applied Thermal Engineering. 2020;**176**(May): 115474. DOI: 10.1016/j.applthermaleng.2020.115474

[41] Ahmad Jamil M, Goraya TS, Ur Rehman A, Yaqoob H, Ikhlaq M, Wakil Shahzad M, et al. A comprehensive design and optimization of an offset strip-fin compact heat exchanger for energy recovery systems. Energy

Conversion and Management. 2022;**14**: 100191. DOI: 10.1016/j.ecmx.2022. 100191

[42] Al Horr Y, Tashtoush B. Experimental analysis of the cooling performance of a fresh air handling unit. AIMS Energy. 2020;**8**(2):299-319. DOI: 10.3934/ENERGY.2020.2.299

[43] Jain JK, Hindoliya DA. Energy saving potential of indirect evaporative cooler under Indian climates. International Journal of Low-Carbon Technology. 2016;**11**(2):193-198. DOI: 10.1093/ijlct/ctt076

[44] Moshari S, Heidarinejad G. Analytical estimation of pressure drop in indirect evaporative coolers for power reduction. Energy and Buildings. 2017; **150**:149-162. DOI: 10.1016/j. enbuild.2017.05.080